The Forbidden Subject

Morse Reese

Copyright © 2020 by Morse Reese
All rights reserved.

Joel Pett Editorial Cartoon used with permission of Joel Pett and the Cartoonist Group. All rights reserved.

No part of this publication may be reproduced, distributed or transmitted in any form or by any means, including photocopying, recording or other electronic or mechanical methods, without the prior written permission of the publisher, except in the case of brief quotations embodied in reviews and certain other non-commercial uses permitted by copyright law.

For permission requests please contact DartFrog Books.

Published 2020
Printed in the United States of America
ISBN 978-1951490577 (print)
ISBN 978-1951490584 (ebook)

Canoe Tree Press
4697 Main Street
Manchester, VT 05255
www.CanoeTreePress.com

DEDICATION

I am dedicating this book to my grandchildren, Bodhi Roamin Reese, Aria Sofia Jackson and Barrett Ryder Reese; and everyone who has been born or will be born in the twenty-first century.

I will probably be dead and gone, but they will live to experience the world we pass to them.

Additionally, I extend this dedication to all living things in the biosphere, both those that survive and those that perish. I will keep working to turn the forbidden subject around. I hope I am not alone and that we are not too late.

WITH DEEP LOVE OF EARTH'S WONDERS,
Morse

CONTENTS

Chapter 1 - About the Author and This Book 7
Chapter 2 - Introduction to the Forbidden Subject 15
Chapter 3 - Planet Earth 19
Chapter 4 - Meditation 31
Chapter 5 - Humans/Us 41
Chapter 6 - Biodiversity 57
Chapter 7 - Immigration 69
Chapter 8 - The Ship Has Already Sailed 77
Chapter 9 - Society 91
Chapter 10 - Controlling the Human Population 103
Chapter 11 - Conclusion 121
Acknowledgments 137
Credits 139

CHAPTER 1

ABOUT THE AUTHOR AND THIS BOOK

Left to right: wife Kathy with dog Heidi and horse Penny; Morse Reese; son Tim; daughter Jenny with dog Molly; son Tony with mule John Boy.

THE FORBIDDEN SUBJECT

Hello, my name is Morse Reese, and this is my first book–as you have probably already guessed by the unusual title for this chapter. I am basically a "country boy" trying to help you, our fantastic planet, and the people and critters living on it by sharing facts and my deeply felt beliefs. While usually someone else writes about the author I have decided to do it myself, because *The Forbidden Subject* is a different kind of book which will ask you to look at yourself as well as the rest of us *Homo sapiens*. Reading about me as I see myself may help you to understand me, see yourself and appreciate all people, whatever their culture, religion, or race may be. You may have to look at compassion, humanity, and other good human attributes in a new and broader way. It will require also looking at many of our unique, and not-so-nice attributes as well.

Additionally, I will challenge you to spend time meditating to truly absorb the content and mold your view of the subject, and maybe even spur yourself into action. We will cover meditating and learn about the connections between the brain and the heart that I learned from Dt. Doty's book *Into the Magic Shop* in an upcoming chapter. As I suspect you have already noticed, this book is meant to be conversational. Despite the seriousness of the forbidden subject, I try to use a light-hearted manner whenever possible. I hope that doing so will make it less forbidden. Getting to know me may also help you to filter the information that I aim to present accurately, knowing that all of us tend to slant information based on our backgrounds and experiences.

I am also asking you to take time while reading to think about yourself–your past experiences and information that you have concerning what is being discussed in this book. During meditation you should take the time to deeply examine your personal, religious, cultural, family, and other values, not only to see if they are sound, but also to evaluate how applying them may affect our future. Your knee-jerk reactions to research information and conclusions may require some deep personal examination before keeping or modifying your final thoughts on the subject. Although they are not by themselves the main subjects of this book, we will be covering our brains and hearts in a new light, and our desire to love, hate, seek revenge, reproduce, and act on our sex drive in a straightforward, basic manner. I would like to take this opportunity to apologize to readers outside of the United States. While I sincerely hope that this book will be available throughout the world, it certainly is heavily weighted to the United States because it is the part of the world that I am most familiar with. It is my strong desire to have this be a book for all of us as one family of humans, and to not have conclusions slanted toward Western values, religions or cultures.

As this is not a novel or a who-done-it, you will not have to wait until the end to learn what the forbidden subject is!

Now about me. I was born in Kingston, Pennsylvania, in 1945 when the Second World War was ending. My father's parents were immigrants from Wales, along with his two brothers. They were a rather poor family living in Scranton, Pennsylvania. My father was a coal miner and later a mine inspector. My mother came from a relatively well-to-do Scranton family. She was a full-time housekeeper raising my older brother and myself in a single-family home in Kingston, Pennsylvania. My dad's income was enough to provide for our basic needs but not enough to provide luxuries. My mom was quite ill during much of her high school years and to the best of my knowledge never graduated. However, due to her family background and friends, she was well educated and quite worldly. My dad was very intelligent, a real leader in high school and an exceptional athlete (football and wrestling) but did not have a college education.

My five-year older brother was a handful for my parents and a challenge for his teachers. He was not academically inclined but was very bright and a natural leader. He built his first car from junkyard parts at the age of fifteen. He loved racing cars and boats and enjoyed helping his friend with motorcycle hill climbs. He became successful at drag racing during his post- teen years until joining the Army. After serving his military duty, he became an exceptionally successful businessperson. He married a girl from England, who was very successful in the hospitality industry. They had two children, one girl and one boy. My brother finished his career as a business advisor and spent his retirement years as a relator, and together with his wife owned and managed a motel. Later he fully retired in Clairmont, California.

I, on the other hand, was a skinny little kid who was usually quite obedient. I had plenty of friends but was not a natural leader. My friends and I got into our fair share of mischief, but nothing serious. However, there was an instance where one of my friends bullied a boy and I joined in. It was verbal abuse, and I have never forgiven myself for taking part. It is interesting how some things stay so strongly with you for all your life. They are usually the things that you are most ashamed of.

I loved the outdoors and was active in boy scouts, hunting, fishing, camping, swimming, canoeing, (We could not afford a motorboat but I loved waterskiing anytime I could bum a tow), hiking, catching turtles, trapping chipmunks, nature study, and outdoor cooking; everything outdoors was great! Nature was amazing and I had great respect and admiration for Native American culture.

My high school years were enjoyable. I was a good student and enjoyed sports; I was good at wrestling and had several friends and girlfriends. I worked summers in an amusement park and discovered I was a hard worker and was very good dealing with customers and fellow employees of all ages.

In college I learned forestry. I was a scared kid, but I quickly made friends who helped me to survive. I was not exceptionally good at academics and had to study hard while earning tuition money at the same time. I joined the Sigma Chi fraternity during the fall term of my freshman year and learned a great deal about other people and myself during my three years as a fraternity brother. My only claim to fame was being a three-time intermural wrestling champion. In all honesty, I only beat some of my opponents because I took the opportunity more seriously and got into exceptional physical condition, making winning during the last period somewhat easy, thus, learning another of life's lessons and something about myself.

During college I was in ROTC (Reserved Officers Training Corp) which helped with tuition and gave me the opportunity to enter the military as an officer instead of an enlistee. These were the Vietnam years and the draft was still in effect, but I was very proud to serve. I was hired as a forester with the United States Forest Service and worked in Michigan's Upper Peninsula from graduation until entering military service in January of 1968. After going through a military engineering school at Fort Belvoir I was assigned to a base in Germany. It was an aircraft repair outfit with many pilots who had served one assignment in Vietnam and were anticipating being sent back. Despite not having been in Vietnam myself, this assignment in Germany exposed me to many of the horrors of military combat and young men dying.

I married Kathy, my college sweetheart, in Germany in 1969. We took a European out from the military in 1970. Then it was back to the Ottawa National Forest in Michigan's Upper Peninsula (the UP). My wife became a teacher and we were transferred to the Nicolet National Forest in Wisconsin. We had three children; Tim, our oldest, was born while we were living in Wisconsin. He was followed by Tony and Jenny, who were born when we were back

THE FORBIDDEN SUBJECT

in the UP. Life in the UP was quite rural, with the nearest standard grocery store about thirty-five miles away and only two snowy TV stations to attempt to see. Winter temperatures were often well below zero and winter work required a snowmobile and snowshoes. It was a great time to focus on family and friends, as there were few distractions. Snowshoe softball on frozen Lake Gogebic and high school sports were main events. Then it was on to the Monongahela National Forest in West Virginia, to be a mountaineer and enjoy the eastern mountains. I took an early retirement from the Forest Service in 1995 and proceeded to take a job as the city administrator for Lewisburg, West Virginia. At the turn of the century I resigned in Lewisburg and moved to Pennsylvania to assist my in-laws living on a small farm homestead near State College, Pennsylvania. We have since lost both of my wife's parents and have built our own home nearby.

Throughout our marriage we have been spending time with children. I usually had at least one child in the neighborhood that I took on outdoor activities. We helped to start a Big Brothers Program in Iron River, Michigan. Kathy has taught elementary children everywhere that we have lived. We hosted crisis intervention children in Wisconsin and had three foster children as well. We have hosted several exchange students, and we host Fresh Air children from New York City most summers.

There you have it–in a nutshell, we love nature, children, and helping people!

> *"Journey with me to a true commitment to our environment. Journey with me to the serenity of leaving to our children a planet in equilibrium."*
> *–Paul Tsongas*

Describe yourself:

CHAPTER 2

INTRODUCTION TO THE FORBIDDEN SUBJECT

You may be surprised to learn that the forbidden subject that this book explores is not forbidden by the government or any other entity; instead, it is each of us; possibly even **YOU**!

<u>Yes-YOU!</u>

<u>No need to feel bad though, as you are joined by most politicians, environmental organizations, religious faiths, charities, family members, neighbors, and friends. I was also included in this, until now!</u>

When you start meditating-yes, I am going to strongly suggest this in an upcoming chapter-you can spend some time thinking about why you and others might not want to talk about: THE HUMAN POPULATION! I suspect that it is largely because it can be an unpleasant subject, which is something that I have experienced when talking with a few people. However, in general I have been pleasantly surprised at how many folks are willing to have a good heartfelt and realistic discussion about the problem and possible solutions. Politicians probably realize that it can be a controversial topic, and they do not want to alienate anyone. They would hate to lose a vote taking on the most important subject to all of humanity, I suppose. Charitable organizations, humanitarian as well as conservation, want to raise contributions and may be afraid that the forbidden subject would hurt donations. They do not realize that their cause may be a lost cause without population control. Religious groups probably have the greatest difficulty with the subject because it may require significant rethinking about their own moral values and those of their organization, as well as the effects of retaining them.

Each of us will need to nurture an appreciation for others' views and beliefs without feeling compelled to convert them to our own thinking, which could create a destructive "we/they" situation.

I personally receive a great deal of political questionnaires. These solicitations often include listings of things that they would like to accomplish, and ask for me to rate them on a number scale from one to twenty or so, in order of importance, with number one being the most important, or to simply check the top three, etc. I have yet to receive a single request that included population control. How can this be when I believe that population control is far and away the most important item for all of humanity and our planet?

This book will be difficult and occasionally unpleasant to read. That is why I have devoted a short chapter to meditation and how our brain communicates with our heart. Understanding ourselves and the crucial role that we must accept to maintain our planet will require serious thought, and a review of existing conditions on our planet and what we have done from our first appearance on earth to today. We will need to base our conclusions on facts as well as past experiences, emotions, and beliefs.

Other chapters will deal with planet earth from creation to today and its interrelationships with the rest of our solar system, emphasizing its relationships with the sun and the moon. We will look at biodiversity and human population statistics with an explanation of how today's population developed, and how to control population. We will consider birth verses death including killing people, immigration, what science has done to create the problem and what it can do to ease the pain, and most importantly, when will too many occur. We will also consider possible actions to deal with the problem, and finally what *you* can do.

I would like to make it clear from the start that while I realize my personal views will creep out despite efforts to suppress them, my goal in writing this book is not to convince you that my beliefs are correct. The goal is to make the forbidden subject a widely considered issue with the realization that dealing with it is more desirable than letting it play out on its own. I also want to make it clear that usually, when I say man or *Homo sapiens* that I am including women within the term. The two are truly equals!

> *"I don't think that it is too hippie to want to clean up the planet, so you don't end up dying of some sort of cancer when you are 45 years old. It enrages me that these big cancer research organizations can't be bothered to man the front lines of environmental protest."*
> —Jello Biafra

CHAPTER 3

PLANET EARTH

I originally wanted to title this chapter *Satellite Earth*, but one of my children was quick to point out that most people think of a satellite as being manmade, so I have decided to back off a little bit from the satellite idea. Years ago, someone came up with the concept of spaceship Earth and it did not stick, as well it should not have, as we cannot control Earth's speed or direction. However, the spaceship concept, as well as that of Earth being a satellite, clearly represents the fact that we are riding on an object that is orbiting the sun, and unlike the space station, we have no backup support. It is up to us to keep it functioning, to provide for our needs of breathable air, drinkable water, food, shelter, and literally millions of other necessary or desirable things.

Thus, throughout this book I will occasionally refer to our planet as satellite Earth to remind us of the reality of being on a very special "nature made", or if you prefer "creator made", satellite. It has the unique ability to support life forms and has done so for millions of years. We must learn to understand it and stop destroying it! While we might like to think of ourselves as caretakers of Earth, the real caretaking that we need to provide is self-control. I will cover that concept and look at changes that we have made in later chapters. For now, let's look at Earth's development and that of the plants and animals that inhabit it.

THE FORBIDDEN SUBJECT

    Let's start by looking at the bigger picture: The Universe. I have great difficulty thinking about the Universe, light years, nanoseconds, and similar things of either great or minuscule magnitude. I understand them, but at the same time I am overwhelmed.

    It is believed that the Universe was created about 4.6 billion years ago. Within it there are believed to be between 100 and 200 billion structures called galaxies. Most of the galaxies are called dwarf galaxies, which are much smaller than some additional large galaxies.

    Our galaxy is called the Milky Way and includes about 100 billion stars. It is smaller than the Andromeda Galaxy, which is often mentioned in books and movies. To get a feeling of its size, it helps if we know that astronomers have examined the Milky Way as far out as five thousand light years. That seems very big to me. When looking at the night sky I can hold up my finger and see the width of it touching two stars. My first thought is that they must be close together, but I know that cannot be true or they would interfere with each other. I also know that our star which we call the sun is 93 million miles from earth and gives us daylight and can burn me if I stand in its rays for very long. I also know that its rays have traveled through Earth's heat shield. Thus, I figure that the two stars must be many times further away from each other than the sun is to earth and I realize that they may not be beside each other, as I see them, but are likely to be vastly far apart in front of or behind each other. All that boggles my mind and challenges the limits of my understanding! The bottom line is our Milky Way galaxy is incredibly large, and it is still tiny compared to the Universe. **Wow!**

    Earth is a planet that orbits the sun. Geologists identify time since the Earth's creation as geologic time, based on the age of rocks and what life forms are found in them. This is because they can estimate time by sampling rocks and using carbon dating to calculate their age. They use the life forms in the fossils to decide when to end one and start another geologic time. Then they apply the name of the rock structure that the fossils they found are in to develop the geologic names for the time periods. Thus, eons, eras, and periods can each be vastly different numbers of years long.

The Hadean Eon is about 800 million years long

The Archean Eon is about 700 million years long

GEOLOGIC TIME	Years Ago (millions)
HADEAN EON - no era - no period - no fossils	4000
ARCHEAN EON - four periods - single celled organisms	3200
PROTEROZOIC EON Eon - three periods - algae and soft bodied invertebrates	2500
PHANEROZOIC EON - three eras:	
PALEOZOIC ERA - seven periods:	
Cambrian Period - earliest fish	544
Ordovician Period - earliest corals	505
Silurian Period - earliest land plants and insects	438
Devonian Period - Earliest amphibians and boney fish	410
Carboniferous Period - two systems:	
Mississippian System - earliest winged insects	360
Pennsylvanian System - earliest reptiles	320
Permian period - no new fossils	286
MESOZOIC ERA - three periods	
Triassic period - earliest dinosaurs and mammals	248
Jurassic period - earliest birds	208
Cretaceous period - earliest flowering plants, dinosaurs in decline	144
CENOZOIC ERA - two periods:	
Tertiary Period - five epochs:	
Paleocene Epoch - earliest large mammals	65
Eocene Epoch - earliest grasses	55
Oligocene Epoch - no new fossils	34
Miocene Epoch - earliest hominids	24
Pliocene Epoch - no new fossils	5
Quaternary Period - two epochs:	
Pleistocene Epoch - earliest humans	1.8
Holocene Epoch - we are living in it!	0

During geologic time there have been five great extinctions. To be a great extinction at least 75 percent of the existing species must be lost. The recorded extinctions are as follows:

444 million years ago at the end of the Ordovician Period: 86% loss.

375 million years ago in the late Devonian Period: 75% loss.

251 million years ago at the end of the Permian Period: 96% loss.

200 million years ago at the end of the Triassic Period: 80% loss.

60 million years ago at the end of the Cretaceous period: 76% loss.

We are presently living in what is expected to be the sixth great extinction. Do you want to be one of the exiting species?

Note that all the extinctions occurred "only" millions of years ago. That is because for most of Earth's existence there were no life forms on land; neither plant nor animal.

Now that we have worked our way up to current time, let's look at an overview of some of the amenities that earth has to offer:

Atmosphere: This serves as a meteor shield, a radiation deflector, a thermal blanket, and a storage area for purified water. It also provides chemical energy.

Water: This is required for all life forms, both plant and animal.

Land: It is nice to have something to stand on, and plants also enjoy the nutrients that it supplies. Additionally, we use many of the minerals and fossil fuels that it has to offer. Plants and animals are found mostly on or near the Earth's surface, in the biosphere.

Core: The interior of the planet is hot molten rock, which feeds volcanos, creates faults, and slowly moves the Earth's surface.

THE ATMOSPHERE

*"Thank God men cannot fly,
and lay waste the sky as well as the earth."*
--Henry David Thoreau (1817-1862), author of Walden

The atmosphere is divided into four major layers. The lower layer is called the <u>troposphere,</u> which is from six to eleven miles thick and contains about 83 percent of the atmosphere. It is also where all atmospheric water is stored, and weather takes place. The troposphere is **not** where global warming occurs. The troposphere's activity is powered by the sun.

The next layer is the <u>stratosphere,</u> which is our "satellite's" heat shield. It extends up to about 30 miles and contains the ozone layer, which reflects most of the sun's radiant energy, keeping us from getting burned up in the daytime. It also serves as a blanket that keeps the heat on Earth at night, preventing us from freezing. <u>It is this ozone layer that is being damaged by chemical contaminants and carbon emissions, which are causing global warming</u>. The changes in this layer have been measured for decades and clearly show the continued increases that cause global warming.

Above the stratosphere is the <u>mesosphere,</u> which extends upward to about 90 miles and contains temperatures more than 3,600 degrees Fahrenheit. It is nice to have the ozone shield below, yes?

The top layer is the <u>thermosphere,</u> in which the international space station is orbiting.

The atmosphere is our chemical lab, working with the sun and surface water. The sun heats surface water, causing it to evaporate and form clouds in the lower atmosphere. This is nature's distillation process creating purified water to rain down upon us. This water cycle is something that you probably were already aware of but is included in this book as a reminder of one of the many miracles of this wonderful planet, which makes it possible for life to exist.

Just a note: The Hubble space telescope orbits about 400 miles above the Earth's surface.

The atmosphere above the troposphere continues up to about 1,500 feet and is mostly Helium. Above that, the atmosphere is mostly Hydrogen.

WATER

Ah yes; life-giving water!
Water on Earth is thought to be distributed as follows: 97.4 percent is found in the oceans and 2.6 percent is what is referred to as free. The "free" water is about 76.7 percent ice, 22.8 percent in ground, and 0.5 percent active. **The water that we constantly use is only 0.5 percent of the 2.6 percent.** "Active" is what I think of as fresh and is distributed as follows: 52 percent in lakes, 36 percent in soil, 7.1 percent in the atmosphere, 3.5 percent in rivers, and 1.4 percent in living things. All the water on Earth is the Hydrosphere, and includes the water on Earth's surface, underground, in the clouds and ice. Water occupies about 90 percent of the Earth's biosphere and covers about 71 percent of the Earth's surface.

The oceans are vast, and we have only explored about 5 percent of them. Most of the oceans are quite deep, averaging 3,700 meters (12,100 feet). Life in the ocean started about three billion years prior to life on land. It is estimated that there are more than two million species of plants and animals living it the oceans, but only 230 thousand of them have been identified. I think we know more about outer space than we do the oceans.

When one looks at all the above percentages, it is amazing how

THE FORBIDDEN SUBJECT

little of the water on Earth is available for us to drink and put to the massive number of uses that we and all of the plants and animals depend on or simply enjoy.

Of course, I guess that the 97.5 percent in the oceans is being used and enjoyed by many plants and animals inhabiting that part of our planet. Our oceans are thought to have been formed in the Hadean Eon. There are five oceans; from the largest to the smallest they include the Pacific, the Atlantic, the Indian, the Southern Antarctic, and the Artic.

The oceans are the source of most of our rainfall, and they determine our climates and wind patterns. Ocean waves, tides, and swells all have tremendous energy in them.

Waves are created by the wind and can be anything from ripples to one hundred feet high. Waves can travel for thousands of miles. A swell is what a wave is called after the wind ceases to push it.

Tides are the rise and fall of the level of the water. They are created by the gravity of the Moon and the Sun, and by the rotation of the Earth. The Moon has more gravitational pull on our oceans than the Sun due to its closer proximity to Earth.

> *"Every drop in the Ocean counts"*
> –Yoko Ono

The oceans were the world's international highways for many centuries, with ships being able to carry large loads over great distances using **wind power**. It was water that was used to transport everything, using streams flowing to rivers, and lakes flowing into more streams, and the great oceans. Boats of all kinds could float downstream using *gravity* as their power source. Civilization began alongside water because it not only provided something to drink and to wash in, but it was also the best method of transportation.

It powered the engines of much of the first manufacturing businesses. Loggers did not even need a boat to transfer their logs to the market; they simply (actually it was not too simple) slid logs into streams, backing water up behind dams and then at the right time breaking the dams to float the logs downstream to the sawmills. We were making things happen without using any fossil fuels, just the sun, wind, gravity and animal power.

Earth's water has not escaped our influence, or should I say degradation. Water in the clouds has been acidified; streams carry giardia, fertilizer, and pesticides, and their water is not safe to drink. Lakes and the fish living in them are contaminated with mercury and other pollutants. Rivers have the same problems as streams, plus several invasive species of plants and animals. Additionally, some rivers have so much water used for farming or other human desires that they are running dry. Marshes and other wetlands are contaminated and losing their native species. The oceans are losing their coral reefs; they have islands of plastic floating on their surface, and more plastic and garbage lying on their bottoms. Groundwater is contaminated and the levels of water in the aquifers are dropping.

Aquifers are vital sources of water and are not quickly replenished by rainfall. Just 6 percent of the groundwater around the Earth is renewed within fifty years. The largest aquifer in the United States is named the Ogallala Aquifer and lies below several states. It was created about ten million years ago by fluvial deposition during the Pliocene Epoch. If drained it would take more than six thousand years to refill it naturally. Today it is being depleted annually by a volume equivalent to eighteen Colorado Rivers. Aquifers lying near oceans are being damaged by saltwater that is drawn into them to fill the voids created by the drop in freshwater levels.

THE FORBIDDEN SUBJECT

> *"Water is more than just another commodity—it is essential to life. Unlike oil, there are no substitutes for it"*
> –Christopher J. Daly

THE EARTH FROM ITS SURFACE DOWN

Despite being about 4.6 billion years old, the Earth has not completely cooled or stabilized. The surface, which is called the **crust**, is a brittle, low-density zone from three to thirty miles thick.

Wildlife, the millions of beautiful animals that entertain us, and provide meat for the carnivores and carbon dioxide for the plants, are found on or slightly below the crust's surface.

Plants, the producers of food, are also found on the surface. Without them there would be **no** food! Additionally, they produce oxygen for us and the other animals to breath. Nice combo–plants make oxygen for animals, which make carbon dioxide for the plants.

In the past the crust has been pushed around, making major changes in Earth's surface and the surrounding oceans. It continues to be moved by tremendous forces in Earth's interior, but extensive changes have not occurred in recent centuries.

The **upper mantel**, which is about 1,600 degrees Fahrenheit and extends down for about 230 miles, contains the cracks, fissures, and plates that have been created by the interior forces. It also provides the escape routes known as volcanoes on the surface, for the release of interior heat, gas, and often molten rock and minerals. There are about 550 volcanoes scattered about the Earth's surface and below the oceans, approximately twenty of which are active at any one time. It is thought that it is quite possible for several volcanic eruptions to occur simultaneously and interconnect, causing much or all the Earth's surface to be covered in ash and the atmosphere to be blackened, so that sunlight could not reach the Earth's surface. Such an occurrence may have been the cause of one or more of the past mass extinctions.

You might wonder why I have been spending time to include this brief and basic geological information in a book about human population. The reason is simple: despite the fact that the main focus of the book is what humans have done and are doing to the Earth, I think that it is good to remember that there are forces in outer space, in the atmosphere, and inside the planet that could exterminate us, and maybe all of Earth's life forms very quickly. There have been five major extinction events in Earth's history. Volcanoes, meteors and significant environmental changes are thought to be likely causes of those past mass extinctions. Earthquakes could also kill off large numbers of humans and other animals, but such destruction is not expected to cause mass extinctions.

Getting back to our look at Earth's interior, the next area is called the **_transitional zone_** and extends for about 370 miles. Below the transitional zone is the **_lower mantel_**, extending another 1,050 miles. This is followed by the molten **_outer core_**, which is about 1,300 miles thick and is thought to be about 9,000 degrees Fahrenheit. Finally, there is the 840-mile **_inner core_**, which is a solid mass.

THE FORBIDDEN SUBJECT

Any thoughts? Write them here:

CHAPTER 4

MEDITATION

Welcome to Chapter 4, Meditation. Please do not skip over this section, as it may prove essential to your understanding of yourself and others.

Thinking and praying are wonderful activities but should be combined with meditation to get a good handle on yourself, as well as the subject of human population and any correcting actions that might be proposed.

In chapter one I referred to the book *Into the Magic Shop* by Dr. James R. Doty, a neurosurgeon. The book's title does not seem to relate to his branch of medicine, but if you read the book you will learn how meditation can be like magic, as well as being a skill to enhance your life.

Dr. Doty explains this while relating the interesting story of his life, and the power and risks of meditation. He breaks the learning process into four steps which he calls tricks. Each one is explained in the book and can also be found online by visiting intothemagicshop.com to listen to an audio version of each exercise. Reading the book will not only provide a deeper understanding of the physiological and mental aspects of the skill but is also very entertaining.

I will provide a quick summary and my take on the subject for the purpose of our discussion, but it will fall short of the book's detail and insight. Trick one is relaxing the body; trick two is taming the mind; trick three is opening the heart; trick four is clarifying your intent. Learning each trick requires practice, and it may take weeks or years to acquire proficiency. The tricks are briefly included at the end of this chapter.

If acquired, the skills will allow you to listen and talk to your brain and heart, both of which are interconnected by neurons that communicate with each other every second of your life. The neurons do not just run from your brain to your heart, but also run from the heart to the brain.

To further entice you to try meditating, let me tell you that getting started is easy and you can try it right now. It is easy because your brain is a virtual chatter box and has been talking almost constantly since you started reading this book. Just stop for a second, to think and listen. Think back to when you started to read this chapter; old turbo tongue was probably talking from the very first word, telling you what you should think and do. It may have said, "This is ridiculous. You don't have the time for such nonsense. Skip to the next chapter." Had you listened to it; you would not be reading this sentence. However, since you are still reading this chapter, maybe your heart overtook the mind and required it to give this poor author a break and hear what he has to say. Think about it, and I bet you can recall what was said then and what is said now. Pause, relax, and think. Can you hear it?

Next time you are in a conversation with someone, think and listen to hear your brains voice; I bet that you will hear it clearly! I am not suggesting that you listen to your brain *instead* of the person talking. What I am suggesting is just the opposite! The listening to the brain should only be a short exercise for you to realize what the brain is constantly doing. Once you can hear the brain, turning down its "quick response" input and focusing on what the person is saying will be easier. You will be able to accomplish this with a sincere desire to comprehend what is being said.

Good listening should include asking questions to clarify and may include asking why the person thinks or feels as they do. These inquiries need to show interest in their thoughts as opposed to challenges. Counterpoints might be best left until you have had time to meditate on what they have said. Other comments are okay if they are not attacking. Rather than using statements, you might try presenting them as, "What do you think" questions to continue the conversation and prevent an argument. The idea is to have a good discussion, not a debate, and **certainly** not an argument. It may take extra effort to listen to the brain and the speaker simultaneously for a short time. However, it should not take long to hear them both. Then you can go back to giving the speaker your full attention knowing that your brain may be working in the background to derail you.

Keep in mind that the brain is the body's spokesperson. It gets its motivation from our body's immediate desires or needs and our life experiences. It can do good things, such as having organs drop adrenaline into our blood and telling us to jump back into the car when we observe a large mountain lion approaching. It can also activate poor judgment and cause us to salivate and gobble down three pieces of delicious cake. Thus, listening to the brain can save us, prevent us from trying new things, or can make us act intelligently.

While the brain is certainly not a devil and the heart may not always be an angel, I like to think about them like some of the old cartoons that I saw as a child. These cartoons had a devil sitting on one of the character's shoulders whispering things in the character's ear to cause trouble. The other shoulder was occupied by an angel who was telling the character not to do it! This concept, while not accurate, helps me to keep things in perspective.

THE FORBIDDEN SUBJECT

Keep listening as you move on to chapter five. You may be able to feel the magic and start to take control, blending thoughts from both the mind and the heart.

The actual quieting of the mind and being able to start talking to your heart directly is difficult but attainable (see trick #3). You should become aware of what your brain is saying and try to further analyze those thoughts, to see if you can develop a broader range of or acceptance of ideas. You do not have to adopt different ideas than yours, but you should consider them and accept the right of other people to have them.

Absorbing and appreciating views that conflict with yours will be required if you are going to truly break out from and discuss the forbidden subject.

Remember: DISCUSS; DO NOT DEBATE OR ARGUE!

The following are Dr. Doty's four "tricks." I am placing all four of them on the following pages, but it is best if you practice each one for several weeks before moving on to the next.

TRICK ONE: RELAXING THE BODY:

a) Find a time and a place to do this exercise so that you will not be interrupted.
b) Do not start if you are already stressed, have other matters distracting you, have consumed alcohol, used recreational drugs, or are tired.
c) Before beginning, sit for a few minutes and just relax. Think of what you wish to accomplish with this exercise. Define your intention.
d) Now close your eyes.

e) Begin by taking three deep breaths in through your nose and slowly out through your mouth. Repeat until you get use to this type of breathing, so that the breathing itself is not distracting you.
f) Once you feel comfortable breathing in this manner, specifically think about how you are sitting and imagine that you are looking at yourself.
g) Now begin focusing on your toes and relax them. Focus on your feet, and relaxing your muscles. Imagine them almost melting away as you continue to breath in and out. Only focus on your toes and feet. When you begin it will be easy to become distracted or to have your thoughts wander. When this happens simply begin again, relaxing the muscles of your toes and feet.
h) Once you have been able to relax your toes and feet, extend the exercise upward, relaxing your calves and thighs.
i) Then, relax the muscles of your abdomen and chest.
j) Think of your spine and relax the muscles all along your spine and up to your shoulders and your neck.
k) Finally, relax the muscles of your face and your scalp.
l) As you are able to extend the relaxation of the muscles of your body, notice that there is a calmness overcoming you, that you feel good. At this point, it is not unusual to feel sleepy or even fall asleep. That is okay. It may take several attempts to get to this point and be able to hold this feeling of being relaxed without falling asleep. Be patient. Be kind to yourself.
m) Now focus on your heart and think of relaxing your heart muscle as you slowly breath in and out. You will find that your heartbeat will slow as your body relaxes and your breath slows.
n) Imagine your body, now completely relaxed, and experience the sense of simply being as you slowly breath in and out. Feel a sense of warmth. Many will feel that they are floating and will be

overcome with a sense of calmness. Continue to slowly breath in and slowly exhale out.

o) *With intention, remember this sense of relaxation, calmness, and warmth.*

p) *Now slowly open your eyes. Sit for a few minutes with your eyes open and just be, with no other intention or thought.*

Breathing and relaxation are the first steps toward training the mind.

TRICK TWO: TAMING THE MIND

a) Once the body is relaxed (trick one) it is time to tame the mind.

b) Begin by focusing on your breath. It is common for thoughts to arise and for you to want to attend to them. Each time this occurs, return your focus to your breathing. Some find that actually thinking of their nostrils and the air entering and exiting helps bring their focus back.

c) Other techniques that assist in decreasing mind wandering are the use of a mantra, which is a word or phrase that is repeated over and over, or focusing on the flame of a candle or another object. This helps avoid giving those wandering thoughts attention. In some practices the teacher gives the mantra to the student who tells no one else the mantra. You can pick whatever word you like as your mantra, or you can focus on a flame or on another object. Find what works for you. Everyone is different.

d) It will take time and effort. Do not be discouraged. It may take a few weeks or even longer before you start seeing the profound effects of a quiet mind. You will not have the same desire to engage emotionally in thoughts that often are negative or distracting. The calmness that you feel from simply relaxing will increase, because

when you are not distracted by internal dialogue, the associated emotional response does not occur. It is this response that has an effect on the rest of the body.
e) Practice this exercise for twenty to thirty minutes each day. The reward for taming the mind is clarity of thought.

TRICK THREE: OPENING THE HEART

a) Relax your body completely (trick one).
b) Once relaxed, focus on your breathing and try to empty your mind completely of all thoughts.
c) When thoughts arise, guide your attention back to your breath.
d) Continue breathing in and out, completely emptying your mind.
e) Now think of the person in your life who has given you unconditional love. Unconditional love is not perfect love or love without hurt or pain. It just means that someone loved you selflessly once or for some time. If you cannot think of someone who loved you unconditionally, you can think of someone in your life that you have loved unconditionally.
f) Sit with the feeling of warmth and contentment that unconditional love brings, while you slowly breathe in and out. Feel the power of unconditional love and how you feel accepted and cared for, even with all of your flaws and imperfections.
g) Think of someone you care for and, with intent, extend unconditional love to that person. Understand that the gift that you are giving them is the same gift that someone gave to you, and will make others feel cared for and protected.
h) As you are giving that unconditional love to one you care for, think again how you feel when you have been given unconditional love and acceptance.

i) Again, reflect on how it feels to be cared for, protected, and loved, regardless of your flaws and imperfections. Think of a person that you know but have neutral feelings for. Now with intention, extend the same unconditional intention to them. As you are embracing that person with love, wish them a happy life with as little suffering as possible. Hold that person in your heart and see their future. See their happiness. Let yourself be bathed in that warm feeling.
j) Now think of someone with whom you have had a difficult relationship or for whom you have negative feelings. Understand that oftentimes one's actions are a manifestation of one's pain. See them as yourself. A flawed, imperfect being who at times struggles and makes mistakes. Think of the person in your own life who gave you unconditional love. Reflect on how that love and acceptance impacted you. Now, give that same unconditional love to that person who is difficult or for whom you have negative feelings.
k) See everyone you meet as a flawed, imperfect being just like you who has made mistakes, taken wrong turns, and at times has hurt others, yet who is struggling and deserves love. With intention, give others unconditional love. In your mind, bathe them with love, warmth, and acceptance. It does not matter what their response is.

What matters is that you have an open heart.
An open heart connects with others, and that changes everything.

TRICK FOUR: CLARIFYING YOUR INTENT

a) Sit in a quiet room and close your eyes.
b) Think of a goal or something that you wish to accomplish. It does not matter that the details of the vision are not fully formed. It is important that such a goal of vision is one that does not involve harm to another or bad intent. While this technique could help you to accomplish such a goal, it will ultimately result in pain and suffering to yourself and make you unhappy.
c) Relax your body completely (trick 1).
d) Once relaxed, focus on your breathing and try to empty your mind completely of all thoughts.
e) When thoughts arise, guide your attention back to your breath.
f) Continue to breathe in and out, completely emptying your mind.
g) Now think of your goal or wish and see yourself as having accomplished it. Sit with the vision as you slowly breathe in and out.
h) Feel the positive emotions associated with accomplishing your goal or having achieved your wish. Experience how good it feels to have taken a thought and turned it into reality. Sit with the positive feelings as you see yourself having accomplished your goal.
i) Once you have seen yourself having accomplished the goal and have sat with the positive feelings, begin to add details to the vision. Exactly how do you look? Where are you? How are people responding to you? Add as much detail to the vision as possible.
j) Repeat one or two times daily (or more) for ten to thirty minutes. Each time begin with the vision of yourself having accomplished your goal. Sit with the feelings. Each time, as you look at the vision add more details. It will start fuzzy, but the more times you do the exercise, the more the vision will become clear.
k) With each time you do the exercise you will find you are refining the vision, as your unconscious mind begins having clarity of the

intent. You may be surprised at what you discover and how you end up achieving your goal. What is important is the goal, not exactly how you get there.

It is with clarity of intent that vision becomes a reality.
HAPPY MEDITATING!

CHAPTER 5

HUMANS/US

This chapter looks at human beings, the strange animals that we are. Before getting started however, I would like to say that when I say man, I mean mankind, which includes both women and men. I will try to rarely differentiate by sex, race, nationality, or any other category that we create. People all over the world seem to be so similar that efforts to categorize are simply unproductive. Often it seems that there are more differences between spouses, family, neighbors, fellow countrymen, political parties, and what have you than between nationalities and races, etc. I also believe that creating we/they situations is destructive.

Looking at humans I see many similarities, some of which I have listed below:

We want to be loved, and we want to love.
Our sex drive is very strong it can give us immense pleasure or cause us to do terrible things.
Most of us want to have a child or children and are willing to make great sacrifices for them.
We want to provide for and nurture our families.
Our greatest pleasure is helping others, but we do too little of it.
We seek revenge.

THE FORBIDDEN SUBJECT

We like to eat.
Our best friends are those with whom we have suffered hardship, and those who we can count on when needed.
We lie too much.
We want to be appreciated.
We will cater to bullies as children and as adults.
We are innovative.
We like to party.
We will turn all inventions into military weapons.
We like to feel that we and those in our group are the greatest.

I hope that you take the time to think, meditate on, and discuss what I have listed. I have included a blank space at the end of this chapter where you can record your beliefs about "us."

Now, let's glance at the good, the bad, and the ugly that makes us what we are. Starting with the ugly, I would like to share with you a story about a hike that my wife and I took one day while visiting a park in Florida.

We were about to set off on a guided hike. Prior to starting, the guide warned us that the wetlands were frequented by the deadliest animal on earth. He said that this animal would kill for sport, hide from its prey, and attack all other animals no matter their size, even when unprovoked. We were instructed to follow his lead, walk closely, and to step off the trail about five feet if he gave the signal. We were then to remain still and quiet. My wife and I knew that there were alligators in the area, but we were not aware that they were so aggressive. Not too long into the hike the signal was given, and we complied with his warnings. Still and ready for anything, we quietly stood then we thought we could hear voices. Sure, enough a young couple walked by. Quickly we realized that the animal he was referring to was people, and it was a reminder to us all that we indeed are the fiercest, most deadly animals, and possess the negative traits that he warned us about. Additionally, we were reminded that we were intruders in the habitat of many other animals, and it was us that needed to be avoided.

Furthermore, the ugly is not restricted to our mistreatment of nature and the environment; it also includes our nauseating mistreatment of other people! Preparing for this book I decided to read *Remembering: Voices of the Holocaust,* by Lyn Smith. The book is a collection of stories as remembered by survivors of the Holocaust. It makes for difficult reading, as it is hard to imagine any human treating another in such horrific ways. Additionally, while reading the book you cannot help but wonder how you would react in any of the multitude of atrocities described. It is my belief that many readers would internally react with "I would not do that." That is certainly what I tended to think. However, upon deeper reflection I have concluded that with the millions of people that were mistreated, and the large number of people who participated in one way or another administering the mistreatment, we have an excellent sampling of true human behavior. Thus, I too may have acted similarly. I am going to share two of the stories that impacted me most heavily.

First is an example of adults following a bully to gain the bully's favor. We all too often think of children as bullies and children as the followers of bullies, excluding ourselves from the mix. Unfortunately, that is not true, and can be realized by knowing that each succeeding German invasion of another country was followed by the non-Jews who were invaded, violently rounding up and torturing Jews to win the favor of the invading troops. It did not matter that the Jews were their former neighbors and friends. They were now just a passage for gaining the desired acceptance. The mistreatment of Jews was often greater in the invaded countries than it was in Germany, and most of the mistreatment was from the locals, not the invading troops.

The second story that I would like to share is of a teenage boy who was captured and taken to a concentration camp. The soldiers routinely bulldozed large trenches outside camps and then marched "excess" prisoners to the trenches for "disposal". The teenage boy in this story was given the job of shooting the prisoners, many of which were entire families, as they lay side by side at the bottom of a trench. He related that he always tried to shoot the parents first as he felt they suffered too badly when they had to experience their children being shot.

Bullies are often very successful people. Hitler was certainly a bully; Teddy Roosevelt was probably a bully, as are several of today's world leaders. We follow bullies. They declare themselves as winners and people want to be on the winning team. It is something that we must constantly be cognizant of.

The latter part of the book included thoughts of survivors looking back at what they had to do to survive, both the suffering and regrets if survival had meant serving the Nazi machine. Many came to the realization that deep religious beliefs often provided the strength to endure.

Enough of the ugly! Let's look at the bad. Remember: the brain is a great rationalizer. It can justify anything you want. It can drive you to exhibiting the traits listed at this chapters start. You might want to take another look at them and think about how they have affected your life and how you can change them within yourself and with others.

Now, let's look at some good. Our ability to love not only ourselves and our mates but also other people, nature, and both wild and domestic animals is exceptionally good. While the ability to love is not exclusive to humans, our brains increase this power and our ability spreads beyond the ability of other animals. Along with love comes compassion. Think of the thousands of charities that we contribute to. Helping people in faraway places, fighting diseases, caring for the elderly, feeding the hungry, helping the homeless: the list can go on and on, and we do these things by giving of our time and money. However, we do not just do these things for the heck of it. We do them because it makes us feel good deep inside. I know that helping others gives me a wonderful feeling.

COMPASSION: THE DEVIL'S TOOL

Wow! it seems like I can say some of the most foolish things. I am a strong supporter of compassion and wish there was a great deal more of it. However, when it comes to population control, I believe that compassion can lead us down a dangerous path, increasing pain and suffering rather than relieving it. Let me explain; as this has caused me many sleepless, tear-filled nights.

Worldwide there are masses of malnourished people riddled with disease and parasites. They are trying to survive at a level of poverty that I can hardly imagine, let alone relate to. Thousands are dying daily. For years there have been many plans to help these people. The plans have included goals to provide the needed food and medical attention. Massive media campaigns have resulted in food donations, increased local agricultural production, and increased medical attention. This all sounds great, so how come the problem has not been corrected, but rather in most cases it has gotten worse. *The Forbidden Subject*, that is how!

While the existing people were being helped, they were also multiplying at an accelerated rate. Pre-teen and teenaged girls were having babies, occasionally four or more before their twenty-first birthdays. The population grew faster than the relief. Providing quality birth control was the missing key needed to prevent the continuation and worsening of the problem. This is one of the realities that may require serious meditation, depending on your present beliefs. The shocking and hard-to-swallow reality may be that letting the population die back to sustainable numbers would have been more humane.

Thus, I have concluded that the real solution must be a blend of compassion coupled with quality birth control. Fortunately, some of today's relief efforts are including birth control. Some of the charities are aggressively attacking the problem, while many are still doing little or are keeping their efforts low key in order to prevent the loss of contributors. Those fears may very well be real. We can change that! We can bring birth control out of the closet and tackle it. We can work together without creating one of the dreaded we/they situations that are so destructive.

The problem of the fear that surrounds the conversation about overpopulation not only applies to relief charities but to almost all other charities and politics. One charity used to say, "No matter what your cause is, it's a lost cause without population control." While this statement may seem to be far-fetched, it is very accurate. Whether the charity is for curing a disease, stopping erosion, keeping plastic out of the water, protecting the loss of farmland, controlling gangs, fighting poverty, stopping global warming, world peace, immigration, saving seashores and coral reefs, endangered species protection, or anything else, the problem can be directly linked to overpopulation.

Looking at cancer, you might ask, "How is that related to overpopulation?" It is, because cancer is mainly a man-made problem. Our advanced science has created all sorts of chemical compounds that have infiltrated most aspects of our lives, including our food, our water, the soil, the air, the climate, our homes, our cleaning supplies, almost everything that I can think of. A large part of the problem is that these compounds are prevalent and interact with each other in ways that simply could not have been tested in the labs. Rachel Carson in her book *Silent Spring* did a wonderful job of bringing this problem to light, which has greatly increased our awareness. However, despite the improvements that we have made, the chemicals are everywhere, and overpopulation drives us to use more and more of them as we lose farmland and try to produce larger crops and more meat from less land.

We use insecticides, herbicides, rodenticides, and so many other chemicals to make our lives easier that there is no avoiding them. Way back when I was in college, one of my professors took a sabbatical to Antarctica to test for the presence of DDT. Amazingly, he found it almost everywhere he looked. It was in the birds and in the eggs the birds laid, and in other animals. This type of evidence and Rachel Carson's excellent work led to stopping most of the use of DDT and many other chemicals. However, we still are turning to chemicals to solve our problems, and it is these chemicals and their interactions that give us cancer.

It is overpopulation that drives us to seek the creation of and use of the chemicals. The lists of carcinogens keep growing yearly as we seek artificial ways of accommodating our growing population. Fighting cancer and looking for cures is probably the largest supported block of charities worldwide. While it is possible to get some cancers from overexposure to the sun, most cancers are the result of issues caused by humans from chemicals that have been created in efforts to support the needs or desires of our ever-increasing population.

Another thing that constantly bothers me is our penchant for war. I cannot blame overpopulation for this, as man has been waging war

THE FORBIDDEN SUBJECT

as far back as the first recorded history and I suspect long before also. Additionally, war has helped to keep our numbers down. Thus, if we were to categorize various actions as they effect human population, war would go into the good column. However, I simply cannot accept war as good any more than I am able to accept compassion as bad. What really gets me about war is that it is something that happens between people who I keep recognizing as very similar, and that it is us, the most intelligent animals on Earth doing it.

I constantly visualize two groups of soldiers in separate foxholes shooting and throwing hand grenades at each other. One foxhole is occupied by Americans of various religions and national backgrounds. They are fighting against Germans, supposedly all the Arian race. My perspective is that the men in the American foxhole are more different from each other than some of them are from the enemy just yards away. Furthermore, they all have loving parents, spouses, children, neighbors, and friends back home. What is the difference? Why are they fighting? The answer is too complex for me.

To make things even more befuddling, after the war is ended the warring parties often become good friends. Wouldn't it have been nice to have done that prior to the war? This was also true for the Japanese and most of our allies. Worldwide there is agreement that war is unacceptable and organizations such as the League of Nations and the United Nations are created to prevent future wars. Everyone agrees that the slaughter of people and the destruction of property, including all kinds of great historical buildings and art treasures, is totally foolish. Yet time marches on and the lessons learned seem to fade away as massive military forces are created at the expense of human needs and the environment. My wife and I have hosted several exchange students and have friends in various countries. We sincerely care about those people nearly as much as we do for our foster and biological children. I have a story to share, which exemplifies this point and is dear to my heart.

It is about Kathy, my wife, and her parents and siblings. After the end of the conflict with Germany they prepared "care" packages and sent them to a family in Germany to help with post-war survival. The family in Germany was very loving and appreciative. When times were more normalized, they exchanged Christmas gifts and stayed in contact with each other for many years. When I was stationed in Germany in the 1960s, Kathy, not yet my wife, and I visited the family. They summoned in the relatives and treated us as royalty. This beat the heck out of war!

Some of the above may seem to be adrift from the forbidden subject, but I am including it because corrective action can best be accomplished with worldwide cooperation. This is particularly important when immigration is included. While immigration in and of itself does not change worldwide population, it is related to extreme overpopulation in certain areas.

With the good, the bad, and the ugly being completed, let's look at ourselves, starting with our first appearance on Earth about two million years ago. I realize that there are differing opinions about how the earth was formed and how mankind was created. For the purpose of this book, it is not important whether you believe in a creator or evolution, as both concepts recognize what an incredibly unusual and wonderful planet Earth is! However, avoiding religion, cultures, and nationalities entirely would not be realistic, as they have often contributed to the building of conceptional walls and barriers to problem solving. Additionally, some of their concepts such as heaven and hell, and angels and devils provide me with words that most readers can easily conceptualize. Again, please excuse the fact that I use examples from Christianity, as it is the religion, I am most familiar with.

THE FORBIDDEN SUBJECT

Please remember that this book is being written by a human being, who is unintentionally burdened with prejudices, and sincerely hopes that you will have the strength to forgive any such weaknesses and focus on the subject. My goal is to change a mentally forbidden subject into an accepted thoughtful conversation about what I perceive to be the single most important subject for all of mankind and this planet. Acting like the preverbal ostrich who buries its head in the sand so that a predator will not see them is how I believe over population is presently being handled. That needs to change.

Solutions will require struggling with our existing beliefs and the absolute need to appreciate other points of view without creating we/they situations. We are humans, one species that needs to appreciate each other and develop solutions. We need to create mutual benefit for both our species *and* the rest of the planet. It is important to keep reminding ourselves that all the plant and animal species got along just fine without us for billions of years and can do so again, but we cannot survive without them. Yes! If we humans should suddenly just disappear, all the plant species and other animal species would probably flourish. However, as amazing as it may seem, if the insects disappeared, we humans could not survive.

So much for feeling important!

Solutions can be found without having to force one person's beliefs upon another. Creating we/they situations or trying to villainize someone else's view is destructive and will undermine our mutual goal. The meditation recommended in this book is grounded in the hope that we can appreciate each other. Meditation should focus on understanding as opposed to looking for ways to change or defeat someone else. The main purpose of this chapter is to recognize our weaknesses and appreciate different opinions as we look at how man has evolved.

HUMANS/US

I fully admit that I am often driven out of control by the overwhelming belief that reducing the human population is essential for the planet's survival as the magnificent thing we have been given, whether by a creator or by natural forces.

Chapter three looked at Earth with the recognition that it is a satellite with an estimated age of four and one half billion (4,500,000,000) years orbiting around a star. Now, let's look at what a short time the two million years that we have been on board this planet is. To "demonstrate" I decided to compare the two times (how long earth has existed and how long man has been on earth) to two distances (one being the circumference of the earth, and the other being the portion of the circumference that represents our time on board). To do so, I drew an imaginary line around the Earth's equator with one end representing Earth's creation and the other end being today. I selected the equator because the Earth is not completely round, and I could find a textbook distance (24,901 miles) for it.

I have made no attempt to calculate any of the ups or downs of this long journey. I have however, selected the top of Mt. Kenya in Kenya, Africa for several reasons. One is that it is close to being on the equator. Another is that it is somewhere other than in the United States. Mostly, I chose it because its peak was pushed up to 5,199 meters (about 17,000 feet) approximately three million years ago, long before humans were on board. Our hypothetical journey will start down the mountain and head east across the Indian Ocean to Sumatra. Then we will go by boat across the China sea to trek across Borneo where we will then board another ship and cross just north of New Guinea in the Pacific Ocean. We will sail close to Line Island and Kiritimati before crossing a vast portion of the Pacific Ocean, passing close to the Galapagos Islands on our way to South America. We will enjoy our multi-cultural journey across northern South America prior to splashing into the Atlantic Ocean.

THE FORBIDDEN SUBJECT

Across the ocean we go, and we will land on the western shores of Africa, but this is not yet the end of our journey. We still must travel through the Congo basin and cross the northern end of Lake Victoria and arrive within sight of Mt. Kenya. But are we back where we started? We are not! We will need to continue traveling until we are about five miles from the foot Mt Kenya. Now that we have established our arrival, let's look at how our population has grown.

For our first about 1,797,000 years on Earth we retained a population of less than one billion. Then, about 218 years ago we started an incredible increase thanks to our ingenuity and scientific advancements.

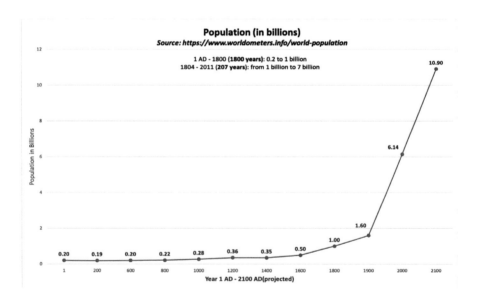

Prior to the population surging, more than 90 percent of Earth's human workforce was laboring away, associated in one way or another with agricultural production and distribution. Children were needed to help with the work and to replace adults, who had a much shorter life expectancy than that of today. Life was more difficult, and it took having several children to get some to adulthood. Waterpower and livestock were the engines of industry. Harvesting was the job of hand labor. Then came the agricultural revolution, which has been so dramatic that today only about 2 percent of our labor force is needed for food production. Shortly after the agricultural revolution started, the industrial revolution began to blossom. Animal power was replaced by ever-increasingly powerful engines. Science was thriving in every aspect of our lives, and we were enjoying a growing life expectancy. Thus, the human population increased rapidly, doubling and redoubling at ever decreasing intervals, bringing us to November 2018, with a population of more than seven and one half **billion** (7,662,293,672) people.

Note the following statistic that illustrates how sudden this problem developed:

<u>**Approximately one sixth (1/6) of all of the people who have ever been born are presently alive and walking on the face of the Earth.**</u>

Should you want to see this phenomenon in real time you can log onto the United Nation's WORLDOMETER where you can watch as our population changes second by second. That website also provides a great deal more demographic information.

One encouraging statistic is as of November 2018, the annual percent increase has fallen to just a little over one percent. While that sounds great–and it is–unfortunately today's one percent is many, many more births than high percentage changes were in earlier years. With today's population being more than seven billion six hundred million, the one percent increase will add more than 77 million people to our planet in one year. The population estimates for the year 2050 are presently 9,771,822,753 and these estimates are based on the continued increase rate dropping to 0.56 percent. A later chapter will cover the significance of such an increase.

I know that I said that I was going to cover the good, bad, and ugly aspects of humans, but I seem to have mostly focused on the bad and the ugly. Maybe that is because most of the big events like wars, exterminations, genocide, mass shootings, mass immigrations, spousal abuse, sexual abuse, and so on get much of the news coverage and thus tend to stick in our minds more. I do, however, fully realize that we as individuals and as various groups and organizations work diligently to do good and humane things constantly for other people, wildlife, and the environment.

On the other hand Wars and actions such as the Holocaust represent some of man's greatest failures. We can do much better!

I again want to encourage meditation in order to fully think about what has thus far been covered and to help prepare yourself for the material that is yet to come. It will also help you to encourage and participate in constructive conversations, if you are so inclined.

What do you think about the human race?

CHAPTER 6

BIODIVERSITY

> "If we are destroying our trees and destroying our environment and hurting animals and hurting one another and all that stuff, there's got to be a very powerful energy to fight that. I think we need more love in the world. We need more kindness, more compassion, more joy, more laughter. I definitely want to contribute to that."
> –Ellen DeGeneres

Planet Earth was discussed in chapter three where we looked at the land, sea, and air individually. Biodiversity goes a step further, looking at how the lifeforms and the environment work together. Various areas of the earth are known as ecosystems, where certain plants and animals live. The total of the ecosystems makes up the biosphere, which is an irregularly shaped thin zone on or near the Earth's surface and throughout its waters. The biosphere recycles its air, water, organisms, and minerals constantly to maintain a wonderfully balanced state. One person added to the definition saying that humans should do their best to imitate it.

As I mentioned earlier, I am a forester. The word biodiversity is a common word in the profession and was used almost daily during my latter years with the Forest Service. Biodiversity simply means the variety and diversity of life forms that occupy one area of land or water. It can be applied to a small area like a yard or pond, or to a large area like a forest, lake, or mountain range, or even to larger areas such as an entire watershed. Biodiversity, while being a relatively easy term to understand, is extremely complex and almost impossible to calculate when looking at a specific area of land. Having to consider all life forms, minerals, elements, and compounds that are present, as well as weather conditions, slope, aspect, and the millions of interactions that could occur, simply cannot be done.

What we do know is that making changes to the ecosystem can have surprising and often very negative consequences. The variety existing in biodiversity has come to mean *good* while the counter term monotype, which means predominately one species, is becoming more and more thought of as *bad*. Monotypes are such things as plantations, orchards, fields of one crop, etc. They eliminate a large part of the diversity, creating a single dominate species. These areas are easier for humans to manage but create large areas where diseases and pests can get out of control, damaging or killing the desired plant species and other species as well. Nature also has created some monotypes, but it has done so over millions of years, creating a biological balance along the way. When faced with the problems created by our monotypes, we try to solve them by applying various chemicals to fight off the pests. Of course, this adds completely new chemical compounds into the ecosystem and the effects are unknown and all too often undesirable. So, biodiversity is having a large variety of plants and animals mixed in the ecosystem, interacting in a balanced, natural way. Recent studies have shown that even intermixing a variety of agricultural crops on the same field can better control pests than the pesticides formerly used.

Insects have shown a surprising ability to adapt and mutate very quickly. Such mutations have made them immune to some insecticides. Thus, some insecticides that worked well to control a specific pest one year might have little to no effect on the same pest a few years later. Then, it is back to the lab to create another pesticide to do the job or find a better and more natural solution. Applying insecticides may not only fail to do the desired job but may also kill valuable insects. Such is believed to be the case with the recent dramatic decline of honeybees. Bees and other insects are the pollinators of the plants, which are the only food producers on earth, thus the bees are vital to the existence of all animal life. That is why it is often said that all the other animals could get along fine without man, but man could not survive without insects. The impact of insects holding such an important position in our lives is particularly striking, since we tend to hate the little critters!

Absolutely everything that we do affects the ecosystem in which the action occurs. Every road, house, airplane flight, invasive species, dam, driving of a car, production of electricity, nuclear bomb, loss of a plant or animal species, yard, garden, island of plastic, pollutant in our rivers and streams, golf course–everything affects one or more ecosystems, and we do not really know all the effects.

> *"The environment is everything that isn't me."*
> –Albert Einstein

> *"When we try to pick out anything by itself, we find it hitched to everything else in the universe."*
> —John Muir

Water is one of the planet's greatest resources, providing for plant and animal life. Rachel Carson keenly made several important points about water and our treatment of it. On page seven of her insightful book *Silent Spring*, she says, "The chemicals to which life

THE FORBIDDEN SUBJECT

is asked to make its adjustment are no longer merely the calcium and silica and copper and all the rest of the minerals washed out of the rocks and carried in rivers to the sea; they are the synthetic creations of man's inventive mind, brewed in his laboratories, and having no counterparts in nature."

On page 39 she enlightens us with the realization that, "By a strange paradox most of the earth's abundant water is not useable for agriculture, industry, or human consumption because of its heavy load of sea salts, and so most of the world's population is either experiencing or is threatened with critical shortages." She continues, "Ever since chemists began to manufacture substances that nature never invented, the problems of water purification have become complex and the danger to users of water has increased." On page forty-four she drives the point home by saying, "Indeed one of the most alarming aspects of chemical pollution of water is the fact that here – in river or lake or reservoir, or for that matter in the glass of water served at your dinner table – are mingled chemicals that no responsible chemist would think of combining in his laboratory." I certainly could not have better described the condition of our water today and it is interesting to realize that the problems associated with it were observable back in 1962. Since then some improvements have been made, but new contaminants have been added and the shortage of potable water has greatly increased. I would bet that even Rachel Carson would not have ever imagined paying two dollars for a twelve-ounce bottle of water.

Invasive species are also an important factor in many ecosystems. I cannot, however, totally blame the problems that they cause on our existing large human population. They are species of plants or animals including diseases that have traveled to ecosystems where they formerly did not exist. Sometimes they were transported intentionally, and sometimes they were unintentionally moved. Some of the transplants blend well into their new environment and are not problem invaders. However, most of the newcomers have caused serious problems, often eradicating a native species. As a forester, it seemed like a new insect or disease appeared every year or two, and our forests suffered dramatically. Some examples of invasive species that you may have heard of or been exposed to are shown below.

MAMMALS
Rabbits, Mongoose, Nutria, and Short Tailed Weasels

WATER DWELLERS
Asian Swamp eel, Northern Snakehead, Sea Lamprey, Crayfish, Zebra Mussels, and several species of Snakes

INSECTS
Spruce Budworm, Gypsy Moth, Emerald Ash Borer, several species of Lady Bugs, Brown Marmorated Stink Bug, Khapre Beetle, Woolly Adelgid Spotted Lanternfly, and the Spotted Wing Drosophila

DISEASES
Dutch elm disease, chronic wasting disease, oak wilt, and Chestnut Blight

PLANTS
Stilt grass, Kudzu, Autumn Olive, Beach Vitex, Giant Hogweed Brazilian Pepper Tree, Multiflora Rose, Giant Hogweed, and Japanese Honeysuckle

Invasive species spread rapidly due to the lack of natural enemies in their new environment, and thus cause massive damage. Annual

THE FORBIDDEN SUBJECT

losses from insects alone just in the United States exceed thirty billion dollars in one year.

Personal experience has exposed me to gypsy moths, which for several years left entire mountain ranges without leaves. Dutch elm disease has almost eliminated American Elm trees from the eastern United States. Chestnut blight knocked the most abundant and marvelous tree along the eastern states down to just a few immature sprouts and some patches of native trees. Emerald Ash Borer is presently killing entire stands of White Ash trees as it continues its spread in the eastern states. The loss of the American Chestnut tree has been particularly hard hitting for me, as it was the most abundant tree in Pennsylvania in the late 1800s and a significant tree throughout the entire east coast. Not only was it abundant, but it was also exceptionally useful and valuable to us for food and lumber. Additionally, it was a mainstay for nut-eating wildlife. The wood was rot resistant, thus it could be used outdoors and stretched for miles as fencing in rural America. The wood was easy to work and very beautiful.

Somewhat surprisingly, much of the wood was riddled with tiny holes; instead of being considered defects, the lumber became known as "wormy chestnut" and was and still is prized for furniture and paneling. People today seek out any kind of old structure, building, cabinetry, or furniture that may contain chestnut wood in order to salvage it for anything from knick-knacks to paneling. The chestnuts were also an important source of income for farmers and rural folks, as truckloads could be relatively easily gathered and sold in cities prior to Christmas. Ever hear the song that mentions "chestnuts roasting on an open fire"? Well, those were American Chestnuts, which are sweeter and better tasting than other chestnuts. Some chestnuts are still sold at Christmas time, but they are European, Chinese, or Japanese Chestnuts, which are larger but not nearly as tasty as the ones now almost extinct.

I said that I cannot exclusively blame the problem of invasive species on the size of our population or on mankind alone. I recognize that some invasive species were transported by other animals or severe weather events, but most were transported by man and the more of us that exist, the greater the chance of such intended or accidental transplants.

When discussing the environment, the subject of extinctions usually comes up. Plant and animal extinctions are of concern because they are indicators of change, which is often caused by man. Extinctions occurred long before man set foot on Earth and were sometimes massive. You may remember from chapter three that some of those extinctions are estimated to have eliminated more than 90 percent of the plants and animals that existed prior to whatever event occurred. There are various theories about what the events may have been, but that knowledge is not vital to the subject of this book. What is important is to realize two things about extinctions: first, that extinctions are possible without our involvement, and second, that we have caused and continue to cause extinctions. Even early *Homo sapiens*, the species that we humans are, caused extinctions. Their ability to create weapons and hunt as teams coupled with the use of fire was too much for the large, slow mammals that formally counted on their size for protection. Fire was not only used in hunting to drive herds of animals or to flush them out of their dens and nests, but it was also used to burn large areas to create or enlarge grassland habitat for the species that man desired. Today our dominance in both sheer numbers and the ability to live throughout the biosphere is causing an alarming increase of extinctions.

One extinction that occurred back in the late 1800s and early 1900s as the agricultural revolution progressed rapidly and the industrial revolution was getting started is particularly interesting and became the eye-opening stimulus to start the environmental movement. It was the extinction of the passenger pigeon. In the 1850s the passenger pigeon was the most abundant bird in North America, and possibly the world. Its numbers were estimated to be between **3 to 5 billion.** The migrations were monumental. The birds flew in fantastically large flocks which numbered in the millions and may have even reached one billion on occasion. When a flock passed overhead the entire sky was darkened and the roar of their wings prevented conversation on the ground, caused livestock to panic, and even caused some folks to think the world was coming to an end. Such a pass over might last for hours. Nesting sites covered as many as 850 square miles, and a single site could be occupied by more than one million birds. Living closely and migrating in large numbers is known as *predator saturation*, which served the birds well. For years people shot the birds out of migrating flocks or even on occasion knocked them out of the air by swinging sticks or throwing rocks. Such harvests killed relatively few birds, allowing the pigeons to maintain their massive numbers. Using predator saturation, the birds were able to go anywhere they wanted without interference or competition from other animals. When the birds flew in, every other animal hightailed it out and headed for cover.

Then came commercial hunting and the predator saturation advantage became the pigeon's major downfall. Newly constructed railroads transported hunters to the nest sites and then carried the tons of killed birds off to market. The telegraph was used to pass on information about nest sites and the locations of migrating flocks. There were no hunting laws so the killing included nets, fire, cannons, and other resources that could not only be applied to flocks but also to nesting areas. The bird's numbers plummeted, and the predator saturation advantage was lost. By the mid-1890s flocks only contained birds in the dozens. Three captive flocks were established but did not successfully reproduce. The last bird was a female named Martha, estimated to be twenty-nine years old. She never laid an egg in captivity and is believed to have died on September 1, 1914.

The tragedy of the passenger pigeon provided stimulus for hunting regulations, wildlife, and habitat management. The concept of conservation was born!

While this book is focused on my concern for the effects of our population on the environmental stability of our planet, I am also aware and appreciate that we have done some wonderful things and continue to do so for all aspects of the environment. The problem is that despite our efforts, our population growth is simply overwhelming the stability of the biosphere.

Most plants and animals are limited to one or a few ecosystems. We humans with our ingenuity are the only animal that can live in all ecosystems and even survive outside of the biosphere.

Understanding biodiversity and the biosphere is important for understanding how the human population affects the Earth, and the realization that the larger our population becomes, the more pressure we will be applying to the environment. We seem to think that we can beat nature, but reality tells us that what we are accomplishing is the destruction of what makes the biosphere work.

> "The earth does not belong to us; we belong to it."
> –Marlee Matlin

Earth is unique because of the presence of life forms. Our measuring of geologic time is based on the appearance of various life forms. So, let's look at what scientists' estimate is the number of life forms that exist and put them into categories that we humans have created.

Total species by %
Arthropods	64.1%
Plants	14.2%
Fungi	4.2%
Mollusks	4.2%
Other inverts	4.0%
Vertebrates	2.7%
Alga	2.4%
Protozoans	2.4%
Nematodes	0.9%
Viruses & Bacteria	0.9%

(Note: Anthropoids have no vertebrates; examples are insects, arachnids, and crustaceans)

The total number of species of plants and animals is estimated between 3 million and 100 million. So far only about 1.7 million have been identified and named. Such a variety creates the possibility of many interactions, leaving a lot for us to learn.

Now that we have some idea of the number of living things estimated to be in the biosphere, let's take the category vertebrates, which are birds, amphibians, reptiles, fish, and mammals, and focus on mammals. Mammals of course are us and the other animals that feed their young with mother's milk through mammary glands.

The following graphic depicts, by weight, the amount (using tons) of mammals that exist on earth. The categories for horses, goats, pigs, sheep, and cattle only include the domesticated ones serving mankind. Wild ones are included in the wild animals category.

BIODIVERSITY

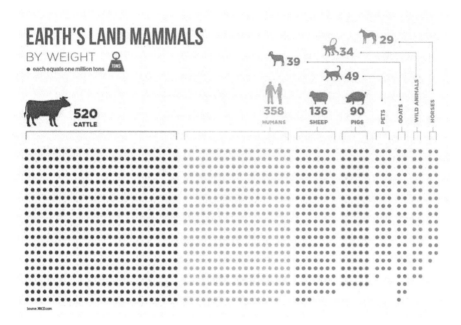

The chart depicts that we and our domesticated animals heavily outweigh the totality of mammals on earth. Domestic horses are the only animals that are outweighed by wild animals.

Now that we have looked at the sheer weight of man and his domesticated animal friends, let's once again look at our population and how it has changed.

For thousands of years our population varied little, fluctuating by only a million or so people from one century to the next. By 1800 our population had crept up to about one billion. But then, it started a steady climb. By 1900 it reached 1.1 billion. By 1930 it reached two billion (an addition of 900 million in only thirty years). By 1970 it hit three billion (that is a whopping one billion more in only 40 years). By 1980 it hit the four billion mark. Another billion and by 1990 we were at five billion people. By the turn of the century (year 2000) we hit six billion. In 2017 we were

THE FORBIDDEN SUBJECT

at 7.6 billion and today, as of 7 am on January 30, 2019, we are at 7,680,606,500.

> "I am pessimistic about the human race because it is too ingenious for its own good. Our approach to nature is to beat it to submission. We would stand a better chance of survival if we accommodate ourselves to the planet and viewed it appreciatively instead of skeptically and dictatorially."
> –E.B. White

**BEFORE MEDITATING - GO ONLINE TO
https://www.worldometers.info/world-population/**
It might take your breath away!

Do you think that we are sharing the Earth properly with other animals?

_____
_____
_____
_____
_____
_____
_____
_____
_____
_____
_____
_____
_____
_____
_____

CHAPTER 7

IMMIGRATION

Immigration is a difficult subject to discuss, because this book is focused on planet Earth being a satellite filled with all of us needing to work together, to keep the planet's amazing biosphere functioning properly. Immigration on the other hand is simply the movement of people from one place to another, and thus, it does not change the Earth's human population.

However, I believe it would be irresponsible not to discuss immigration, because the reality is that we do not presently function on Earth as one people. Instead we are **divided** into separate countries, with each country's population being heavily affected by immigration. Great countries have been built on immigration, and nothing in this book is intended to demean individuals of any race, nationality, or religion. Population planning needs to consider the influx (immigration) and loss (emigration) of people due to migration. Should one country want to decrease their population, any person that moves to another country is a step in the desired direction for them and a step backward for the receiving country if they too want to achieve a population reduction. Not only are the countries affected in general, but so are all their citizens. I say this because migrants affect economic stability, as they require food, water, housing, education, employment, medical care, etc.

Additionally, each immigrant may mean one less baby may be born to an existing citizen to attain a population goal or to maintain stability. Some migrants will contribute productively to the new country's economy, while others will be a drain on society. The immigrants also are likely to have their own babies after arriving in the new country, which again affects the number of children that may be born to established citizens. Statistics indicate that the immigrants will have, on average, more children than established citizens, further reducing the number of children that may be born to the established citizens. Each year's population change will be:

$$births + immigrants - emigrants - deaths = change\ in\ population.$$

Since starting to write this book, immigration and its problems have risen to the forefront of discussion throughout our planet's societies. What is happening is mostly not discussion, but heated argument and assigning blame. Somehow folks seem to overlook the core of the problem and instead focus on the results. The creation of the we/they that I so greatly dislike is occurring, and the assignment of negative attributes to people whose opinions conflict with one's own beliefs is rampant. What riles me so much is my belief that both sides really want the same thing: **TO HELP THE SUFFERING INDIVIDUALS.** The arguments should change to discussions focused on determining the root of the problem and possible constructive solutions. Of course, getting a small group to agree on possible solutions is far easier than getting various countries to agree on the solutions. Implementing them is still a greater task, but one with more chance of succeeding than the bickering. The "discussions" are occurring both within countries and between countries. I believe that Brexit in Great Britain is largely because of immigration issues within and between countries, as is the wall that President Trump wants to build, and sanctuary cities in the United States.

Immigrants have many reasons for relocating, but mass migration is usually an attempt to flee from undesirable conditions or to attain a better future, which usually means finding quality employment and a safe place to live freely.

This chapter is highly focused on the United States because I am not well versed on conditions in other countries. I have left a blank page at the end of the chapter for you to use to record your thoughts and/or what you know about where you live.

When thinking about immigration, my first thoughts turn to the United States because it is founded on immigrants. The Statue of Liberty displays the inscription, "Give me your tired, your poor, your huddled masses yearning to breathe free, the wretched refuse of your teeming shore. Send these homeless, tempest – toss to me, I lift my lamp beside the golden door."

The Declaration of Independence includes, "We hold these truths to be self-evident, that all men are created equal, that they are endowed by their Creator with certain Unalienable Rights, that among these are Life, Liberty and the Pursuit of Happiness."

Those two writings are from years ago, before overpopulation was an issue. They remind us that some of the reasons for giving up almost everything you have and hoping for a better future are due to social and economic issues in the homeland and can be powerful forces. Times were different then; America was a vast land open for exploitation at the expense of its native inhabitants, and a government of the people by the people and for the people was a fresh new concept.

I am going to sidestep from the core of what I am specifically writing about to mention that the United States has not always been as pure as we Americans might like to think. This paragraph

THE FORBIDDEN SUBJECT

is meant to humble, as the treatment of the Native Americans was nothing less than disgraceful. The forced immigration of Africans and our using them as slaves is another example of a deplorable violation of our declaration of all men being created equal. Also, our immigration practices have not always been compassionate. Prior to the start of World War II, a boat load of Jewish children headed for our shore. They were sent to us by their parents in Germany to save them from the Holocaust, but we rejected the boat, which had to return to Germany delivering them to the wrath of the Nazis. I believe that this enforcement of our immigration laws differs from conditions today but is worth knowing.

Today, this country and many others have populations that are straining our economies and the environment that provides everything for all of us. All people must work together to be compassionate by solving the core of problems worldwide. Producing more and more people is not the solution and overstuffing one nation to relieve another will not work.

In today's world immigration, both legal and illegal, has developed into a very controversial subject. Masses of people are risking their lives, and some are losing their lives, in efforts to flee their homes and find safe livable places for their families. Different opinions abound in the countries that the immigrants are attempting to enter. Laws are being broken and the we/they finger pointing that I hate so much is preventing cooperative development of quality solutions. In the United States, one group who wants to violate existing laws and allow unlimited immigration is framing their position as compassionate and they paint the position of others who want to enforce the laws and work for other long-term solutions as inhumane, uncaring, self-centered and barbaric. Worldwide, politicians are being replaced. Immigration policies and laws are being challenged and disobeyed. In the United States *"sanctuary cities*" are being created in defiance of Federal law. In Great Britain the *Brexit* vote was largely driven by immigration issues.

Another immigration issue that is greatly affecting population growth in the United States is a portion of the 14th amendment to the U.S. Constitution which declares, "All parties born or naturalized in the United States, and subject to the jurisdiction thereof, are citizens of the United States and of the state wherein they reside." What a logical and appropriate item to include in the amendment at a time when immigration meant a several month's boat trip across an ocean and hard work to find or build housing and earn a living. Today however, a pregnant woman can board a plane in a foreign country and fly to the United States in a few hours, just days before the expected delivery date, and a new U.S. citizen is created. This is what is called "birthright citizenship," and the new citizen becomes an "anchor baby," which coupled with immigration laws opens the door for the new citizen's parents, siblings, and adults of the extended family to legally migrate to the United States. Not quite what the writers of the 14th amendment had in mind! But as we all know, times change, and in this case, I believe that now is the time to change this amendment and associated immigration laws.

While I am sure that solutions to immigration can be found if all concerned would cooperate, immigration is not the focus of this book. I have included it because it is an indicator of and contributor to the overpopulation of individual countries and other political subdivisions and must be considered when drafting population plans.

Immigration includes both legal and illegal immigrants. I mentioned illegal immigrants in several of the above paragraphs and would like to mention a fact and my personal thoughts about illegal immigrants here. It is estimated that there are presently twelve million illegal aliens in the United States. While being here may be somewhat better than where they came from, I think that it is an unacceptably lousy way of life. Their status prevents them from joining the society that they sought. Imagine having to hide out being in constant fear of being caught and deported and living in a manner that makes it almost impossible to earn a quality income. Additionally, illegal immigrants are easy targets for businesses that want to take advantage of them and gangs that want to recruit them. I cannot imagine that this is the dream they had when they left home. Also, their presence in the new country is occupying one human spot that might otherwise be taken by someone that is legally applying to enter and is being a detriment to the legal society which they are not able to join. While it is euphoric to think it possible, I believe the best solution would be to improve conditions in the homeland, removing their desperate need to leave.

As a side note: The organization Negative Population Growth Inc. has issued a paper titled <u>FORMULATING IMMIGRATION POLICY IN A TIME OF REFORM: A PLAN FOR THE COMPENSATED REPATRIATION OF ILLEGAL ALIENS</u>. It provides some provocative reading for citizens of the United States.

Remembering my formula:

Births + Immigrants − Emigrants − Deaths = Population Change

You might think that immigrants would be a small amount of the population change as compared to the births, but not so for the United States. The following graph shows how the U.S. grew its population by births from 1970 to 2010 and by immigration for the same period. It indicates that immigration slightly exceeded the growth by births. The chart then was extended to estimate, based on existing trends, how each would increase by the year 2050.

IMMIGRATION

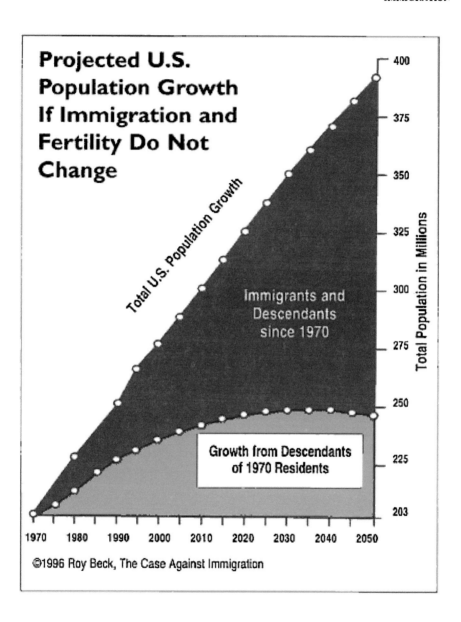

WOW! Births are expected to level off with immigration tripling. Can the environment handle it? You may want to come back to

THE FORBIDDEN SUBJECT

that question after completing this book.

Now, enjoy your meditation and jot down your thoughts about migration on the following blank space.

What do you think about immigration?

_____
_____
_____
_____
_____
_____
_____
_____
_____
_____
_____
_____
_____
_____
_____
_____
_____
_____
_____
_____
_____
_____
_____
_____

CHAPTER 8

THE SHIP HAS ALREADY SAILED

Before starting to write this book, I decided not to predict when the world's population would exceed more than what the planet could endure. I knew that past books had made predictions and various readers spent their time debating and disagreeing concerning the predictions about years, numbers, and effects. There was a tendency for the public to lack interest because overpopulation was not going to happen in their lifetime, and science would take care of any problems. I remember reading that some group thought that an ideal world population should be about one hundred thousand, while others thought that the population could keep doubling and not only "we" could handle it, but it would be a good thing. Debates focused on details and avoided the main problems.

Well, as you can probably conclude from the title of this chapter, I have changed my mind dramatically. I have concluded that if there is to be any debate, it should not be when it *will* happen, but rather when it *did* happen.

First let's look at the numbers. It does not matter whether you think our existing seven billion plus is too high or if you think eight billion would be okay, or ten billion or twenty billion; I hope that we can all agree that increases cannot go on forever. Such growth is like a chain letter, which is great at the start but will crash in the future.

Despite what I just said I am going to include a look at a final population number and various predictions of when it will be reached.

THE FORBIDDEN SUBJECT

Scientists have determined that a world population of **ten billion** would almost fully utilize the earth for the survival of mankind. Livestock would be a luxury that could not exist because meat and dairy production is inefficient, since many pounds of vegetation must be consumed to produce one pound of milk or meat. Grazing land would have to be converted to vegetative production to feed us. Now, if you want to argue about the ten billion figure go ahead, but I do not think there is a lot of wiggle room. The concept is the important part to grasp, not the number itself.

> *"It's coming home to roost over the next 50 years or so. It's not just climate change; it's sheer space, places to grow food for this enormous horde. Either we limit our population growth, or the natural world will do it for us, and the natural world is doing it right now."*
> –David Attenborough

Now, let's look at when it is predicted the ten billion will be reached. That year is up for "discussion" and varies from 2080 to 2120 depending on estimated birth rates. The year 2000 is a good mid estimate and is based on little change to our present growth rate. If correct, it will not occur in a future generation but will be a gift from us to **today's children!**

Whether we will ever make it to ten billion or not depends on what we do today and any of many catastrophes that could occur. I will leave most of those possibilities to the entertainment industry. My position is that <u>the ship has already sailed</u> and our existing population, which is more than seven and one half billion, exceeds what is **sustainable**. I say sustainable because I believe in working with the biosphere rather than trying to outsmart it.

"Our globe is under dramatic environmental pressure: our globe is warming, our ice is melting, our glaciers are rescinding, our coral is dying, our soils are eroding, our water tables are falling, our fisheries are being depleted, our remaining rain forests are shrinking. Something is very very wrong with our ecosystem."
—Richard Lamn

The question should be how to stop increasing our population, and then how fast to reduce it. The people at **Negative Population Growth** recommend reducing our population growth slowly, and then reducing our population slowly thereafter. The timetable for accomplishing this would be about 2030 for having the growth stopped and then no target date for the reduction. Since 2030 is almost ten years away, we have time for study and discussion to develop a plan for reducing our numbers. The key is the goal of this book: *to start caring conversations about the forbidden subject.*

Later I will explain the basis for my belief, and then I will welcome you to decide for yourself based on your beliefs. Right now, I would like to take a quick look at plausible events that could cause major extinctions, possibly eliminating us from the planet and make this entire book moot.

Remembering back to chapter 3 *Satellite Earth*, we learned that archeologists have determined there have been five mass extinctions of plants and animals between the Permian and Triassic Periods. One extinction is estimated to have eliminated 96 percent of all life forms. While I do not think that anyone knows for sure what caused them, it is obvious that things presently beyond our influence could happen and vastly reduce our numbers or completely kill us off. The most likely are geological (volcanic) and astronomical (meteor) events.

THE FORBIDDEN SUBJECT

Volcanic would be a series of interconnected category eight volcanic eruptions, leaving most of Earth in the dark due to a sky filled with ash and the ground covered with a thick layer of ash. These two conditions would prevent plant growth, and no plants means no food.

Should one or more large meteors penetrate our atmospheric shield, their impact on the planet's surface could be almost as serious as the volcanic eruptions. Earthquakes could also be serious, but expectations are that they would not be interconnected or occur simultaneously, thus they should not cause widespread extinctions. Slow steady movements of the Earth's surface are constantly occurring but are also not expected to cause sudden and major changes to our planet.

Unfortunately, there is a new threat to add to the list. It is **us**; it is **nuclear war!**

I will not say much about this because it would be too complex to speculate, but man's power has risen to a point that can rival some of the most destructive powers of nature. Should there be survivors, their exposure to radiation may leave them sterile, to live out their lives as did the last passenger pigeons. It is up to us to ensure that this does not happen.

Let's return to why I believe our present population is too large, but first let me say why it is difficult for me, and possibly for you, to say that we already have too many people on board. Since my life may be vastly different from yours, you must take the description of my environment and extrapolate it to your life. I live a comfortable life, which allows me to enjoy the benefits of modern life in America and overlook Earth's problems. I live in a nice house with heating and cooling systems, so I am always comfortable. Food is varied and plentiful. I have a great family, dogs, equine, and plenty of land on which to ride, hike, hunt, and just plain enjoy the great outdoors. Life for me is good, therefore I should have nothing to be concerned about.

But I am concerned. So, I will now open my eyes to observe and relate what I can see, and think about problems that I have learned exist on planet Earth. The following pages list many of the indicators of overpopulation and biosphere damage that I see and believe to be very real.

INDICATORS OF OVERPOPULATION AND BIOSPHERE DAMAGE

Refugee Camps–largest camp 886,778 people. There are 60 babies born daily. It is estimated that more than 48,000 babies will be born in camps this year; the average stay of residents is 17 years.

150 million homeless people

Street gangs

Aquifers are drying up

Acid rain

Extinctions

Millions of people without health care

Global warming

5,000 dying daily due to starvation or other malnutrition problems–most are five years old or younger

Endangered species

Mercury-laden fish

Young girls having multiple babies during their teen years

Families needing to have both parents working, leaving children with inadequate supervision and little time to play, socialize, and learn with their parents

New Diseases to attack humans, other animals, and plants

Invasive species

THE FORBIDDEN SUBJECT

Air quality indexes

Polluted water

Rivers and streams running brown after rainstorms

Fertilizer and pesticides in our rivers and streams

Wells going dry

$2.00 for a bottle for drinking water

Rivers without water because it is used by us before flowing to its confluence

People wearing face masks to walk on city streets

Radioactive areas

Clear mountain streams with water unsafe to drink

Excessive deer populations

Cities running out of potable water

Iceberg algae

Deteriorating and loss of coral reefs

Islands of plastic in our oceans (8 million tons being added each year)

Losing polar ice

Thawing of the permafrost

1,300 Super Fund Sites (hazard waste) in the U.S.

Cancer – the manmade killer

Illegal immigrants

Police in schools

Increased security checks

Turning away migrants in dire need

The loss of the American Chestnut

Almost ½ of the world's population living in poverty

750 million people lacking clean drinking water

1.6 billion people without appropriate shelter

Wow, that is a lot of stuff! I cannot believe that if this world was produced by a creator that they would be pleased with our changes. Likewise, if the world was created and crafted over billions of years by Mother Nature, I believe "she" also would be appalled by what we have done.

While typing all the scatterings of problems above caused largely by overpopulation, I started to realize how many of them are occurring right here on my land and in my small rural community. They include:

Acid rain, global warming, endangered species, extinctions, invasive species, mercury-laden fish, new diseases, air quality indexes, polluted water, deer ticks, Lyme disease, streams running brown, giardia in the streams, overpopulation of deer, police in schools, illegal immigrants, and increased security checks.

ELABORATIONS

Endangered species: Several endangered or threatened species of birds visit my property, and I am presently cooperating with state and federal agencies on a project designed to prevent the Golden Winged Warbler from being added to one of the lists. On a good note, Bald Eagles are making a comeback in the area. Thank you, Rachel Carson!

Extinctions from the State: Mountain lions and wolves no longer inhabit my land and losing them has contributed to the creation of a large deer population. The deer cause serious problems, including the reduction of the forest's biodiversity.

Stream quality: As a youngster, drinking water from a mountain stream was fine. Today I would not drink from any stream due to giardia.

Virginia White Tailed Deer: This is a further explanation of extinctions above. Deer were a part of the fauna of Pennsylvania's woods. The word Pennsylvania is a combination of Penn for our founder William Penn, and sylva, which means forest or in the forest. Thus, Penn's forest = Pennsylvania.

The Native Americans, lions, wolves, and perhaps other animals killed and consumed deer prior to the white man's arrival. White man also liked to eat venison and make things from their body parts. Combined with harvesting almost the entire forest landscape in the late 1800s and early 1900s, and the uncontrolled fires that followed, the deer were almost eliminated. Then came pastures, cornfields, hay fields, young forest regeneration, gardens, landscaping, orchards, vegetable production, and hunting regulations. It became a deer eutopia. Deer returned and their population soared.

Deer graze in warm weather and browse (eat leaves, bark buds, and soft woody material) in the winter. The joke is, that the deer will eat anything you want to grow and will not touch the plants you do not want. As a forester I can tell you that they have seriously changed (damaged) the forests. Their winter browsing consumes any part of tree seedlings that stick above the snow. Thus, good tasting species were greatly reduced and only a few species did well. Such consumption greatly reduced biodiversity and has set the forests up for major trouble when specific insect pests and diseases arrive.

We have now looked at existing conditions on Earth that I think indicate an existing overpopulation and have considered major events that might cause mass extinctions. Now I want to extend the above information about deer and look at how their populations fluctuate. Doing so might give us some idea of what might happen to us if there is no catastrophic event and we continue to allow our population to grow. I realize that we are not deer and vary greatly from them, but how nature deals with their overpopulation may be enlightening.

I have chosen deer for several reasons: (1) I have been involved with deer and their management throughout my forestry career. (2) They have a short lifespan. (3) My father-in-law was the professor at Pennsylvania State University in charge of the deer pens and experiments with the deer to aid the Game Commission with their management. (4) Deer in the wild have relatively short lives, usually less than five- and one-half years. (5) My contact with them in Michigan and Wisconsin (sometimes referred to as the Northwoods) is at the northern extreme of their range, where they are exposed to severe winter conditions. In captivity they may live for ten or more years, but in the wild they rarely live past five- and one-half years. (The age of deer is usually referred to in half years, such as two- and one-half years, because the fawns are born in the spring and hunting season is in the fall. Harvested deer are fawns or one and one half, two and one half, etc. years old.) As explained above, deer adapt well to man. During the summer, even in the Northwoods, they have excellent conditions for growth.

Winter is different story. The severity of the North Woods winters causes them to yard up (bunch together in groups). They usually select a cedar or other stand of evergreens. The evergreens intercept much of the snowfall, making walking easier and allowing them to lie together in small groups to share body heat. When browsing in the winter, they eat everything edible on ground level and everything as high as they can reach; after eating what they can while standing on all four feet they stand on their back feet, reaching as high as possible. Thus, the larger deer can get more to eat, leaving the fawns to suffer. A heavy snowfall might weigh branches down into their reach, but these will be consumed quickly. Their browsing activity

THE FORBIDDEN SUBJECT

creates a portion of the woods void of small vegetation, and the level where the vegetation appears is referred to as the browse line.

Food is scarce and desperation causes them to dig to get vegetation below the snow, and even to eat soil. They seem to know when conditions are extra bad to just lay down and stay down, as the energy expended to find food is greater than what would be obtained walking and digging. When I worked in the winter, I entered the forested area by snowmobile which left a compacted snow track which the deer would then use to venture out of their yard. Unless the winter was mild, some of the herd would die. Causes included starvation, freezing, diseases that spread rapidly in the close conditions, and fighting.

I do not know if this deer story applies to us, but it is nature's way, and nature may apply it to us if we do not manage ourselves.

Leaving the winter conditions and returning to Pennsylvania, deer are now dying from CWD (chronic wasting disease), a new disease discovered several years ago. The disease is not only killing deer, but also other ungulates such as elk and moose. Overpopulation – new disease?

We humans also now have new diseases to contend with. How many? How recently? Why? What do you think?

After my thinking that I had finished writing this book, I received an excellent Forum Paper from Negative Population Growth, Inc. titled *Looking Ahead*, written by Peter Seidel.

When I originally wrote this chapter and titled it "The Ship Has Already Sailed"; I had reached that conclusion based on the long list of what I believed were indicators of overpopulation and biosphere damage. Mr. Seidel's paper has added authoritative data in addition to my noted changes that are taking place on our earth. His findings have reinforced my position of the ship having already sailed, and I will share some of them with you. He points out that humans tend to believe what we want, often putting rational thinking and scientific findings aside, refusing to accept that we are creating problems. He has observed that we tend to live in a "here now" manner and are not good at looking to the future and results that "living for today" will heap upon future generations. "We don't notice that <u>today</u> there are 232,000 more people on our planet than yesterday; 68,000 more acres of arable land have been seriously degraded or abandoned from agriculture; 35,000 more acres of forest have been obliterated; desertification has claimed over 2.5 square miles of land in China); and water tables around the world have dropped further." Yes, these are changes that take place in one day! Did you notice them?

> *"We are using resources as if we had two planet earths, not one. There can be no 'plan B' because there is no 'planet B.'"*
> *–Ban Ki-Moon*

It seems that Moon was thinking ahead of the footprint network!

The Global Footprint Network, www.footprintnetwork.org, estimates it would take 1.7 planet Earths to produce what we now consume and to absorb the carbon dioxide emissions that we create. Even more shocking, they estimate that if everyone lived the way Americans do, it would require 5.1 planet Earths to live sustainably.

The affluent do not tend to see or feel the plight of the poor or realize how similar the poor are to the weaker deer in my earlier story. People are dying daily because they cannot survive the living conditions in their "human yards," such as refugee camps and deprived villages.

The following graph from the Ecological Footprint Network shows the gap between our demands on nature, which they call our **ecological footprint,** and nature's capacity to fill our demands, which is referred to as **biological capacity.**

THE FORBIDDEN SUBJECT

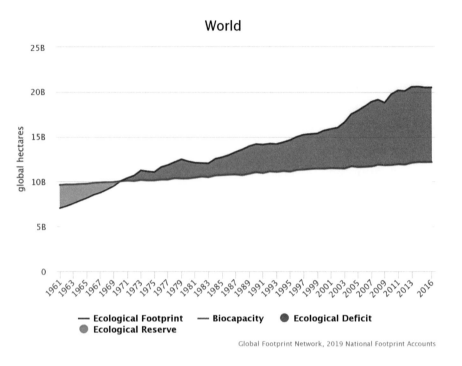

The graph indicates that the biological capacity of Earth was exceeded in about 1970 and has continued to get worse each year. Thus, my position that the ship has already sailed is now supported by some research.

> "Buy land, they ain't making any more of the stuff."
> –Will Rogers

Will Rogers knew then that land was valuable, and we now must realize that there will not be any 1.7 planet Earths to serve our existing demands and continue leaving nearly one half of our population without adequate resources. Certainly, there will not be any 5.1 planet earths any time soon to allow bringing our existing population up to the living standards enjoyed by the average American.

It appears that we are presently out of control. We must create sustainability, which will either require a vast reduction of our lifestyle, our population or a combination of the two.

According to a recent Gallup polls Americans are less concerned now about most environmental issues today than they were in 1989. Are you one of them?

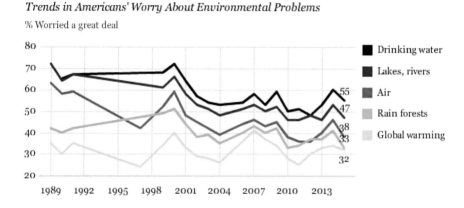

Trends in Americans' Worry About Environmental Problems

"Man has lost the capacity to foresee and to forestall. He will end by destroying the earth."
–Albert Schweitzer

We can continue to make Mr. Schweitzer correct, or we can change the forbidden subject from forbidden to top priority, and act to reverse our error!

This chapter hopefully provided you with some material for meditating. I hope that you will agree with my belief that our present population is too large. However, if you do not, the book's goal is to start conversation, not to convince you that my beliefs are correct.

THE FORBIDDEN SUBJECT

What is your population goal and why?

CHAPTER 9

SOCIETY

"We won't have a society if we destroy the environment."
–Margaret Mead

Chapter 5 looked at humans as individuals and focused on how alike we all are, and how we need to realize that we are really one people riding on a satellite called Earth. Now we are going to look at today's reality of different societies, which combined with geographic location have created different countries. Societies are not unique to people; many animals and even plants may have their own societies. I will look at animals that vary greatly in social contact, but to some extent do so within their species and even between species.

However, first I would like to share the words of John Muir, who recommended taking the time to live with nature for a while in order to better see and understand life and society.

"Keep close to nature's heart—and break clear away, once in
a while, and climb a mountain or spend a week in the woods.
Wash your spirit clean."
–John Muir

Earlier I related the story of the passenger pigeon and the importance of its society, relating that the loss of its society was probably the cause of its extinction. Individual, small groups of the pigeons were cared for in captivity but did not breed and thus did not survive. Bees and ants are known for their societies, where they not only live together in communal homes that they build, but they each have specific jobs to perform. Birds fly in flocks, fish swim in schools; there are herds of horses, colonies of prairie dogs, etc. The animals communicate with each other and even communicate between species, whether intentionally or not. One example is when a chipmunk sees a hawk flying overhead it gives a warning call. That call is not only heard by other chipmunks, but by all the nearby animals, many of which have learned the meaning of the call and respond accordingly. My knowledge of such interactions is limited, but I know that scientists are constantly learning that communication between animals is more evolved than we originally thought.

Before completely leaving nature, let's look at some individual species and the differences between them and us. I am going to start with mating. It seems to be a very strong desire for most species. Disputes over who mates with who can lead to death battles, so sex drive is strong. Caring for their young is also deeply ingrained into many animals. I have seen small birds individually and in groups take on a large predator to protect their young. Pairing up as mates seems to be varied. Geese are believed to mate for life and will leave a migrating flock to stay together if one of them is sick or injured.

For other birds, infidelity is more common than we originally thought. Studies have shown a clutch of eggs to have been fertilized by two or more males. Most animals will nurture their young again, as they are willing to risk their life to fight off an assailant. Then, on the other hand, some like the cow bird are so laissez-faire that they go to other birds' nests, kick out some of the eggs, lay their own, and let the other bird sit on the eggs and raise their young. Turtles lay their eggs, walk away, and leave them to hatch; then the newborns must fend for themselves.

Animals, like coyotes and baboons, hunt together as a group or as a team. Each individual has a specific job to outwit their prey.

Now, on to human society. Human societies are complex. The greatest differences are likely **not** our physical appearance or structure, but what our *brains* believe. Our beliefs have been developed over the centuries, and we have developed different cultures, different norms for different countries, ethnicities, religions, races, and even different ways of life and thinking in different cities, towns, villages, and even families. The Revolutionary War and the Civil War in the United States are two excellent examples of how different individuals in the same family can differ in their thinking so strongly that they chose opposite sides during war.

Our ability to build things has added to this complexity by giving us inventions, so that we can live in any and all environments.

I know little about population planning because I have never seen a population plan, and I think that I can say the same for most political leaders. I am not aware of any state or subdivision thereof having one, and to the best of my knowledge the United States lacks such planning. Thus, this chapter is going to focus on planning commissions and zoning, subjects that I was deeply involved with during my five years serving as the City Administrator for Lewisburg, West Virginia.

Lewisburg is a small city of about four thousand people; as such, I wore several hats as its administrator. I was the zoning officer and supervised the building inspector. I worked closely with the Planning Commission, as well as businesses and individuals applying for permits. It was a wonderful educational experience, hence I learned about the workings of government, the people that it serves, and the interrelationships that they have.

THE FORBIDDEN SUBJECT

Most political subdivisions of the United States have zoning ordinances administered by planning commissions, which may largely be the same folks who drafted the plans. These plans should more accurately be called development plans, as they are **heavily growth oriented!** Communities that tried to stop growth were required to change such plans. The direction was to create plans showing areas for expansion of housing, business, manufacturing, transportation, etc. "We must grow, or we will die" was the generally accepted view nationwide. There was no consideration of the environment and human needs such as scenery, culture, recreation, and socialization.

Most of what I am about to say is taken from a great book *Better Not Bigger* by Eben Fodor. My copy is dated 1999, and though it is slightly dated, everything I read seems to be accurate for today. The book should be required reading for anyone in government at all levels and planning commissions. The book includes twelve myths, which I have abbreviated below:

MYTHS

1. *Growth provides needed tax revenue.*
2. *We have to grow to provide jobs.*
3. *Limiting growth will cause housing prices to rise.*
4. *Environmental protection hurts the economy.*
5. *Growth is inevitable.*
6. *If you do not like growth, there is something wrong with you.*
7. *Most people do not support **growth management** or environmental protection.*
   (Note: **growth management** may mean increasing it or reversing it. Recommended actions would depend on existing natural and

SOCIETY

man-made resources, and the ability and desire of the affected society to increase or reduce them.)
8. *We have to grow or die.*
9. *Vacant or undeveloped land is just going to waste.*
10. *Visual preferences are not important.*
11. *Environmentalists are just another special interest.*
12. *There is no such thing as the public interest.*

Such myths have led us astray and fueled the fire of population growth.

You can see from his list that Fodor believes in environmental protection, and that he does not believe population growth is needed for a quality community or economic stability. Should any of the listed myths puzzle you, read the book–his positions are well founded and supported by sound data. The trick is to improve **quality** of life **not** the **quantity** of people.

When planning you need to consider the full growth spectrum. What that means is that every additional person, house, business or whatever affects the community, and may require additional construction or manpower costs for the community. Planning needs to consider traffic, air, water, solid waste disposal, open space, electric power, forests, hospitals, ambulance service, minerals, schools, day care, cultural and recreational facilities, scenery, police, fire protection and all other government responsibilities.

In 1996, the cost of providing one new house was estimated to cost **the community** $24,500. Such a calculation is difficult as you may be able to add one house with no extra social cost, particularly if your existing facilities are oversized. However, at some time you will have to add a new wing on the school, construct a new landfill, widen a road, and the list continues. Each addition may reduce some valuable social attributes of your community such as open space, scenic views, valuable farmland, nature preserves, etc. Thus, you may have gotten away with adding one house and having the only loss be the blocked view that the neighbor had to suffer, but when you need a new wing on the school the price tag will likely be several million dollars. That is the concept, and it is very real.

The taxes received from the new home will not cover the added social expenses. Thus, taxes will need to be raised for the entire community to cover the difference. That is called "publicly subsidized growth." Growth may create new jobs, but it usually increases unemployment, as other people moving into the community usually fill the better paying jobs. Taxes go up, crime goes up, traffic slows down, etc. The only improved facet of life created by growth is diversity. Growth will bring in new interesting people, new churches, a variety of restaurants, expanded choices of food in the grocery stores, etc.

The above information about community planning is important because it is directly related to population growth.

Growth at some time must be limited, as we live in a finite world with finite resources!

Let's get out of your wallet and look at the environment. It was included as one of the myths, which thought that environmental protection hurt the economy.

In 1994, the Institute for Southern Studies ranked the fifty United States into two categories. One list was for environmental quality of states, referred to as "green states," and the other was for economic health of states, referred to as "gold states." Each list was put in order with the best in the category being #1 and the worst being #50. The two scores were added for each state and then two new lists of ten each were made. One list was for the top ten states (lowest combined scores) and the second list was for the ten worst states (highest combined scores) The lists showed each state's score for gold and for green. **The results are:**

	Top 10 States				Bottom 10 States		
Rank	State	Gold	Green	Rank	State	Gold	Green
1	Vermont	3	1	41	Arkansas	47	37
2	Hawaii	1	4	42	Indiana	38	47
3	New Hampshire	6	2	43	Kentucky	45	40
4	Minnesota	2	7	44	South Carolina	44	42
5	Wisconsin	9	6	45	Tennessee	41	48
6	Colorado	11	5	46	Texas	40	49
7	Oregon	8	9	47	Alabama	46	46
8	Massachusetts	12	8	48	Mississippi	49	43
9	Connecticut	4	18	49	West Virginia	48	45
10	Maryland	10	12	50	Louisiana	50	50

As you can see from the chart, environment and economy are not contrary to each other, but go hand in hand. Taking care of the environment boosts the economy. The study shows that was the case in 1994, and it will be even more important in the future, because everything we have is built from natural resources. No plants = no food; no resources = no water, no homes, no cars, no internet, etc.

> "We have lived by the assumption that what was good for us would be good for the world. We have been wrong. We must change our lives, so that it will be possible to live by the contrary assumption that what is good for the world will be good for us. And that requires that we make the effort to know the world and to learn what is good for it. We must learn to cooperate in its processes, and to yield to its limits. But even more important, we must learn to acknowledge the creation is full of mystery; we will never clearly understand it. We must abandon arrogance

and stand in awe. We must recover the sense of the majesty of the creation, and the ability to be worshipful in its presence. For it is only on the condition of humility and reverence before the world that our species will be able to remain in it."
–Wendell Berry, Recollected essays 1965-1980

As a forester, my colleagues and I frequently talked about sustainable forestry, which is managing the forest so that it can produce water, wood, recreation, wildlife, and its many other benefits, forever. That is not protecting it as wilderness but managing it as a living, dynamic community of life. Trees could grow providing their many benefits to the air, water, soil, wildlife and recreation for people. Then at maturity they provide wood and paper for our use. Following the timber harvest a new stand of trees would be established to start the cycle again. Sustainability should also be applied to the planning of our communities, nations, and world. *The present human population and the direction that it is heading is simply not sustainable.*

"Growth for the sake of growth is the ideology of the cancer cell."
–Edward Abby

From an individual community standpoint, the Catch-22 of growth applies. It goes like this: *the better you make your community, the more people will want to live there, until it is no better than any other community.* The trend to re-locate to improve your situation has been going on since history began. People will only stop moving there when the overall conditions are no better there than in any other town. When you set up the platform for your Christmas train set–yes, I know that this activity is no longer standard for today's children–you lay out homes, churches, stores,

SOCIETY

farms, animals, mountains, trees, etc. Imagine if each Christmas you kept adding to the scene. Would the cluttered mess be very attractive? Well, you might say "of course not, the platform is too small, we just need to add another sheet of plywood and we will have a beautiful layout again." True for your train set-up, but not true for satellite Earth. Our planet is finite. The concept of sustainable communities is not different; the idea is to build great places to live and not clutter them up or destroy the farms, open areas, and forests to expand the cities. Then you will have sustainable communities located on a sustainable planet.

The U.S. Congress supposedly recognized such, when they signed the U.S. National Environmental Policy Act of 1970 with its Section 101, which reads:

> *"The Congress, realizing the profound impact of man's activity on the interrelations of all components of the natural environment, particularly the profound influences of population growth, high-density urbanization, industrial expansion, resource exploitation, and new expanding technological advances, and recognizing further the critical importance of restoring and maintaining environmental quality to the overall welfare and development of man, declares that it is the continuing policy of the Federal Government, in cooperation with State and Local Governments, and other concerned public and private organizations, to use all practical means and measures, including financial and technical assistance, in a manner calculated to foster and promote the general welfare, to create and maintain conditions under which man and nature can exist in productive harmony, and fulfill the social, economic, and other requirements of present and future generations of Americans."*

THE FORBIDDEN SUBJECT

It would be nice if our Congressional people would read Section 101 every now and then and take it seriously. Please note the word **restoring**, which indicates a realization that back in 1970 there was environmental damage that needed correcting. Do not get me wrong–some very good things have happened, but I suspect if an inventory could be made it would reveal that the environmental direction that we have moved is in many aspects, backward.

We in the United States and places elsewhere have not only failed to learn how to build sustainable cities, but we have also built almost instant towns and cities that have grown into depressed areas and ghost towns. That is what happens when a community is not based on a good inventory of its resources and is instead driven by few or even one nonrenewable resource. Such resources are usually either newly found resources for which there is a rapid growing demand, or newly found sources of treasured resources. Examples of the first category are coal, oil, and gas, with examples of the latter category being gold, silver and diamonds. The existence of money, paper currency, and more recently electronic currency has helped to fuel the growth of such cities and has also made it possible for other cities to invent attractions that are not based on any natural resources.

So, let's build a non-sustainable city. It will be in a dry, desolate area where it requires many acres to sustain even one head of cattle. Water is scarce, as is food. Oil, which has been created over millions of years below the surface of our sample area, is discovered. The oil can be refined into fuels to power the newly invented motors that are driving the machines of manufacturing, military might, and overall development. Money makes it possible to purchase the resources that the area lacks, and who cares that it takes millions of years to make any more oil?

Thus, the foundation of the new city is not sustainable, so the

SOCIETY

city is not based on stability. I could keep going with the description of how this imagined city grows and dies but I feel quite sure that you already can clearly visualize it, because you have probably observed several such places which have suffered economically, damaged the environment, left dilapidated monstrosities, and created hazard waste sites.

Now for the city with no special natural resources, only man's ingenuity, which has built something that is very attractive and entertaining. How is it going to fair the long run? I am thinking about places with gambling, racing, amusement parks, etc. and I really do not know how long they will last. Communities that have developed on natural attractions such as lakes and beaches have done well for centuries, so maybe with wise **controlled** development these man-made attractions may have a chance for survival. I need to point out that while I said in my example that the one "instant city" had no special natural resources, everywhere on earth has beauty and natural resources that exist. If ingenuity is used to focus on rather than damage what is there, combined with the man-made attractions and wise limited growth, survival may be possible.

Thus far in this chapter I have been looking mostly at individual societies and their sustainability, and how we are adversely affecting nature and the environment. Now, I would like to look briefly at all of us as one society and how we are affecting ourselves. To do this, I am using information from an article titled, "There is No Planet B" that appeared in the *Nutrition Health Letter* dated October of 2018. It was written by Sam Myers, a physician and the principal research scientist at the Harvard T.H. Chan School of Public Health. He is also the director of the Planetary Health Alliance, and he was responding to an interview about expectations for our future health.

He painted a rather bleak picture for the very reasons that we have been looking at when discussing planet Earth and the environment. He noted that humans have changed just about all the natural systems on Earth and have done so in an unprecedented short period of about 60 years. He noted that the changes have been caused by population growth and our individual increased use of resources. When ask if we have not become healthier in that same time period, he agreed that worldwide we have increased life expectancy and reduced child mortality. He noted that, "The problem is the growth is not sustainable; we are mortgaging the health of future generations to make our lives better now." He went on to explain that the reduced air and water quality, infectious diseases, and environmental changes such as drought, floods, and violent storms ultimately will affect every dimension of our health.

Once again, my thoughts return to the overcrowded deer yard, refugee camps, and all the new diseases that have appeared in recent years. Children born today, not the hypothetical future generations, will have to bear the burden of our greed and lack of self-control.

What do you think about society and zoning?

CHAPTER 10

CONTROLLING THE HUMAN POPULATION

SEX: THE GREAT EQUALIZER

The formula for immigration was a simple one.

births + immigrants − emigrants − deaths = population change

The same formula holds true in this chapter, except this chapter is looking at Earth as a whole. Thus, our new formula is simpler, as it does not include immigration or emigration:

births − deaths = population change

It is probably this chapter dealing with sex and birth control that has contributed most to making population control *The Forbidden Subject*.

It is a simple formula, but if the intention is to reduce our population, solutions are difficult. Had we acted years ago; the solutions would have been more acceptable. Continuing to wait will only leave horrid options and failing to act may mean our extinction. Corrective action even now will be extremely unpleasant. Acting will require great cooperation and challenge every moral fiber. We

THE FORBIDDEN SUBJECT

will each be working from our personal standards, which will be different. Standards are moral, religious, cultural, and personal, and will need to be adjusted or tolerated to save our societies, and possibly our species. A great deal of calm and self-control is required. Solutions will need acceptance of other people's views and values. Trying to force ours onto others is a formula for failure.

The goal must be clear, and it is **not** *for* **you** *or* **me** *to be a winner, it is for* **us** *to successfully reduce our population, eliminate hunger, eliminate poverty, and improve the environment.*

The history of controlling the human population is very short, if we are talking about intentional control by people. Only recently some nations have acted.

Reducing our population can and must be done by us, or *other forces* will do it, and the *other forces* will likely be far less pleasant than what we can accomplish cooperatively!

Thus, if we desire a reduction, we must have fewer births, more deaths, or a combination of the two. Let's first look at the fewer births option, which requires birth control and opens the door for disagreement. Talking about birth control will tend to provoke controversy. Thus, I must now plunge into treacherous waters and ask readers who have beliefs against birth control to give very deep thought to it. Are you going to let those beliefs prevent accomplishing the goal stated above? Times have changed, as has the condition of our planet. A creator or someone who said that they were acting for a creator may have given such direction, but it was men that recorded it as they perceived it. I do not believe that they could have possibly imagined the conditions of today. Birth control was not needed because for centuries our population was relatively low and steady. Only recently has a sudden surge created the problem, **which is destroying what was created.**

To start the birth control conversation, I will address this chapter's subtitle, "Sex, The Great Equalizer." Back in the old west, as the U.S. was being settled, there were men of various physical statures, with the big strong men being dominate. Then came the six-shooter, allowing a small man to gun down a brute. Thus, the weak became equal to the strong. So how does that relate to sex?

As the six-shooter then equalized physical strength, both then and now, sex equalizes financial strength. Sexual desire is a powerful, sometimes overwhelming desire and pleasure that humans of all financial statuses can equally enjoy or abuse. Thus, the impoverished are on par with the billionaires. The folks in the refugee camp can have the same screaming, kick a hole in the wall sex as the wealthy. I use this concept coupled with knowing that the sex drive in animals is also strong and can even lead to deadly confrontations. The subtitle and somewhat crude explanation may seem superfluous but are key to my position that abstinence from sex and the rhythm method have not proven successful. Using them has caused increased teen and other unwanted pregnancies. In other words, they have failed. Thus, I am not considering them as feasible methods of birth control, and I am going to assume that we as a species are going to continue having sexual intercourse. Thus, we must have and use methods of birth control that work.

While only women can, at present, have a baby, it takes both a man and a woman for conception. Both can enjoy the sex, and both are responsible for the results. It is not just the woman's job.

Before getting into specifics of birth control, let's look at existing reality. Birth control methods require knowledge about what is available and planning *prior* to having sex. Additionally, they may require professional medical help and they cost money. Consideration needs to be given to people of all economic classes, but it is the poor that require the most help, as they can least afford

THE FORBIDDEN SUBJECT

to have children and tend to have more children than those of us who are better off financially.

Reality Check: Almost one half (more than three billion) of the world's population lives below the poverty level, having an income of less than $2.50 per day. Eighty percent of our population make less than ten dollars per day. This does not provide them with the funds to purchase birth control pills, patches, morning after pills, or even condoms. They certainly cannot afford to have children, but they do, at high rates. Edouard Balgbedor, the UNICEF representative in Cox's Bazar, Bangladesh, has reported about 60 births PER DAY in the refugee Camps and Informal Settlements. That takes us back to my "great equalizer" comment, leading to – for enjoyment, what else can they afford to do?

While the poor need tremendous help, others of us who may be better off financially are also part of the population problem and need to control the number of children that we have. After all, our children will consume far more resources than children living in poverty.

Worldwide the condom is still the most frequently used method, but it requires teamwork between both partners in the very heated and emotional seconds prior to intercourse. The condom is less expensive than other birth control devices and has the advantage of providing some protection from sexually transmitted diseases. However, today there are several other options, and the medical community is working to develop and test even more options for men and women. Statistically speaking, people with financial stability have fewer children, both from less desire to do so and the ability to afford any of the many options of birth control now available, and with access to the medical advice about how to properly use them.

The most effective methods of birth control are operations for men or women to eliminate their ability to conceive. Medically, the vasectomy, which does not always work, is the operation for the man, and it is easier and safer than medical alterations for women. Research is looking at other medical options which would be less risky for women. Details are not needed for this book, but the bottom line is that husbands and wives have medical options, as do single men and women.

Birth control methods and various forms of protection from sexually transmitted diseases are changing so rapidly that I can only recommend that if you are interested talk with your doctor.

Should you be interested in birth control and what several countries have done or are doing, that information is available online on several sites. At least three countries are experiencing success. However, you are unlikely to enjoy hearing what they have done or are doing. As I have said several times before, the easy solutions were available years ago, but may be lost now! Fortunately, there is one bright note out there; some of the best success has come from increased education of women. Not just medical education, but overall education is critical, as is having their society accept them as equals with men and respecting their rights as human beings. These educated women are often opting to have fewer children so they can have quality employment and the time to do things in addition to caring for the family. Yes, some of these women may be using birth control methods that you may not approve of, but it is working for them. Having better choices for birth control would be great, and medical researchers are working to develop them.

Again, this is not a how to book; it is a want to book. So, let's get back to wanting to reduce our population. Rather than looking at massive numbers covering the entire world, I am going to focus on each of us as individuals, particularly the woman during her reproductive years. Over the long run, limiting ourselves to having two children could stop population growth, but it would take a long time. If a couple would limit themselves to only two children that

THE FORBIDDEN SUBJECT

would not prevent grandchildren or great grandchildren prior to their death. Thus, before their death they may have contributed to the addition of several more than two people to our population, depending on the number of children their children have. Knowing this, I have selected 1 ½ as a goal number of children to have in order to reduce the population. Somehow, even though I know it is just an average, having 1 ½ babies causes me to chuckle as I cannot help but visualize the one child standing next to their one-half sibling. The bottom line is that limiting ourselves to an **average** of 1 ½ children per woman should stop our population from growing and will start a slow decrease, which would prevent serious demographic problems.

Reminder: The above is looking at the entire Earth, not a specific country. To look at a specific country you must add immigration and emigration to the calculation, leaving the possibility that reducing new births to 1 ½ may not stop population growth for a specific country; such countries must also control immigration!

We will now move on to death, which will include natural deaths and the killing of people. Killing people; yes, killing people. Remember, to reduce our population we must get the number of deaths to exceed the number of births. Thus, if we cannot do it with birth control, we will have to increase the death portion of the equation.

Prior to recent efforts our population was mainly controlled by people being killed either by nature or by ourselves. At this point I am going to assume that killing is not your first choice for corrective action. Thus, when we get to looking at corrective actions, we will need to focus mainly on reducing births. The next few paragraphs, however, will briefly look at history when killing was working..

The following are some of the _bad things_ that are beneficial for controlling population growth, thus _have been good_ at keeping our population in check:

Bad is Good

WAR	ABORTION	DROWNING
GENOCIDE	SUICIDE	EARTHQUAKES
MURDER	FLOODS	FREEZING TO DEATH
DEADLY DISEASE	HURRICANES	
STARVATION	TORNADOS	

Now, to look at the opposite; some of the _good things_ that _have been bad_ due to allowing our population to get out of control:

Good is Bad

PEACE EFFORTS	CURING DISEASE	EXTENDING LIFE EXPECTANCY
FEEDING THE STARVING	SCIENTIFIC ADVANCEMENTS	

I hope that we can make our future brighter by keeping and improving the good things and reducing or eliminating the bad things. Our efforts should be focused on attaining population control while increasing the good and reducing the bad, as we achieve the overriding goal of moving our population toward whatever we decide to be best for our homeland.

Surprisingly, looking at present conditions we may still be doing a better job of killing people than we are of birth control.

Killing is a more complex subject than you would think. We

kill people all the time. I did not say murder; I said kill. Let's look, excluding war:

Annual Deaths by killing people
Auto accidents: 1,240,000
Cancer (the manmade disease): 9,600,000
Suicides: 800,000
Shootings: 250,000
Accidents, excluding auto: 161,374
Starvation: 8,565,820

That totals 20,617,194, which is a lot of killing!

To examine killing, let's start way back when man first appeared on earth. There were animals then, which had been killing other animals for centuries. The food chain thrived – plants make food – animals called herbivores eat plants – animals called carnivores eat meat (other animals) – and some animals called omnivores eat meat and plants. When we came along, we joined the omnivore group. Animals that ate meat were also called predators, and the animals that they consumed were prey. The killing of the prey was often very brutal, but that was nature's way. Humans also often killed their prey in brutal ways but evolved, inventing more humane ways of getting the job done.

Thus, killing has been a way of life since life first existed.

I have left the killing due to war and genocide mostly out of this book because they vary so greatly across time, and despite being positive effects on reducing the human population, they are two things that should be eliminated.

We live in a vastly unbalanced society from a financial standpoint. Very few people hold a tremendously large portion of the world's wealth. There are reportedly 2,208 billionaires, and they possess a collective worth of 9.1 trillion dollars. The richest have about 112 billion. On the other hand, nearly ½ of our population – more than 3 billion people – are living in poverty.

Thus, if someone would come up with the idea of simply killing people to reduce the population, who should we start with? Well the wealthy, some who may also think of themselves as important, might suggest starting with the poorest and then work up. Is that the right choice, or should we start at the top and work down? If you start with the poorest you reduce the world's population by one and do little for the environment. Choosing the wealthiest first and taking their money, which they will no longer need, would also only reduce the world's population by one but will provide 1,192,000 checks of $100,000.00 each to be handed to the poor. Additionally, the wealthiest person was consuming large amounts of the Earth's resources each year, so the environment would get a nice little boost while more than two million people, due to the recipients being part of a family, are exiting the ranks of the impoverished. I guess the above is my way of pointing out the imbalance of wealth existing on our satellite.

The human brain is interesting when talking about killing. Should one person murder another person we get very upset and try to capture the murderer and bring them to justice, which might be life in prison or killing them. On the other hand, if one of "our" generals orders 40,000 troops into battle knowing that more than 15,000 of them are estimated to be killed, with the hope of killing more than 20,000 of the enemy, that is okay.

Furthermore, if the battle is won, the general becomes a hero; parades are held in their honor and they might even be elected to be the leader of their country.

(Side note: The enemy soldiers might have been just as fine human beings, as were the soldiers on our side of the battle).

Likewise, if a transportation committee is asked to increase the speed limit on a highway and the data that they are provided indicates that doing so will result in 18 more deaths annually, the consumption of an additional 12,000 gallons of fuel and 87 percent of the public favors the increase. Despite the predicted loss of lives, and fuel consumption; the committee is likely to approve it.

What the above scenarios indicate to me is that it is okay to kill people if you do not know which individuals are going to die. If you know who is to die, it then can be murder. What interesting things our brains are.

Other killing includes suicide, abortion, pulling the plug on a suffering loved one, and execution. These killings are usually not considered murder but can be controversial!

I have had some personal experience and opinions about such killing and suspect that you may have also.

I enjoy hunting, so I have killed, dressed out, and butchered wild animals. Additionally, I have raised domestic animals and slaughtered them for food; and have "put down" pets to relieve their suffering. One time was exceptionally heart- wrenching for me. I have had several dogs throughout the years, but one was a special companion named Daisy. Daisy was a basset hound. I have a special love of basset hounds and have raised several, and even got to enjoy a litter of pups once. Each dog had different dispositions and skill levels.

Daisy however was quite different from all my other bassets. Bassets are usually hunters, but not Daisy. She was an adult when my wife rescued her from the local animal shelter, and I do not have any idea of her early life experiences, but she had no interest in chasing anything. She learned exceptionally rapidly, whether it was leash training or working in the woods. If I was moving, she kept moving. If I stopped, she stopped and evaluated the situation, lying down close by if I was using a hand tool, and lying well away at the sight of the chain saw. She would howl for hours if I left home without saying good-bye and telling her "Stay." No one staying at home could get her to stop the howling; she was beside herself being left unknowingly.

Daisy was a friend to the wild animals. I do not know why she was that way, and I have little idea as to how the animals knew it, but they did. The first time that we came upon a flock of turkeys she walked out among them causing no disturbance. Had any of my other dogs done that, the turkeys would have been running and flying in all directions. Of course, other dogs would not have been walking. Another time when my wife and I were eating dinner outside our camper with Daisy sleeping nearby, a doe and her fawn ambled out of the woods and grazed alongside Daisy. Daisy raised her head and watched but made no other move, and the twosome grazed for a while, then left.

Side note: This was the same doe that continually attacks one of our other dogs because that dog went to check out her fawn when it was a newborn.

Unfortunately, Daisy developed cancer on the roof of her mouth, and despite treatment the cancer spread into her head, causing pain and preventing normal movement. She needed to be put down and our veterinarian would do it and take care of her remains, but I knew that I had to do it. There was no way that I could have anyone else do it, and I could not leave her alone when she needed me the most. I had to be with her when she died.

I have related this personal story knowing full well that a dog is not a person, but this dog was similarly close to me and the story expresses how killing can be done with love. **However, I agree with Will Rogers who once said, "If there are no dogs in heaven, when I die, I want to go to where they went."**

You may have noticed that I have included abortion under the

THE FORBIDDEN SUBJECT

killing category. I have done so because I believe it is killing, and I also believe that if we are going to cooperatively discuss population control, we should not try to apply soft terms to difficult subjects. However, if you think that I completely oppose abortion you would be wrong. I think that it is killing, and I believe that we should try to reduce it greatly, but I think the formula for doing that should be:

no unwanted pregnancy = almost no abortions

In the case of an unwanted pregnancy, I only see three options: have the unwanted baby, give the unwanted baby up for adoption, or have an abortion. None of these options are desirable, and each could turn out to be a big mistake, or it could be the right thing to do for the individual. The decision is compounded by the fact that many people will be affected by it, not just the fetus and the potential parents. Additionally, the parents can be quite young; what they think and feel presently is likely to be different from what they will feel in the future. I fully believe that it is the people most involved and affected by the decision who should be making it. Whenever there is a subject, often a moral one, where there is significant disagreement, each party needs to be careful. Particularly when the difference is near a 50/50 split, they need to ask themselves: "Should I force my opinion upon my fellow citizens if I would be devastated if the tide might turn leading to a decision forcing me to do what I believe to be wrong." To not mince words, in the case of abortion, should they support a law prohibiting abortion if they would be devastated by a law establishing criteria that would require an abortion? We are going to need a great deal of mutual respect and we must develop a tolerance for different ideas if we are going to constructively turn population growth into population reduction. Plans to correct the overpopulation problem are likely to require many undesirable actions. What I said earlier and is worth repeating is "the ship has already sailed", and with it the easy decisions and actions. What is left may be difficult to stomach; what will exist if we postpone action to the future will be worse, and lack of action will likely lead to the loss of human society, and possibly even to our extinction.

The abortion issue gives me an excellent opportunity to elaborate on what I mean when I express a disliking for, we/they situations. Two groups have emerged in the United States. They have assigned names to themselves that are designed to reflect favorably on their positions. The group names are pro-life and pro-choice. Pro-life is opposed to abortion, focused on saving the life of the fetus by having the government prohibit abortion. Pro-choice feels that the possible parents and their families are more familiar with the situation and are thus better suited to make the choice between the three difficult alternatives mentioned earlier.

The issue has become very heated and has created a greater divide between two political parties. The Democratic Party has taken on the cause of pro-choice while the Republican party has aligned with pro-life. Can you truly believe that everyone in a particular political party thinks one way and vice versa for the other party?

Individuals within each party are not free to waver from the party's position; the issue is now "set in stone" and has become a win/lose situation. I have not heard of any effort to identify the base problem or to develop alternative solutions. Cooperation and collaboration are not considered.

Thus far I have not heard anyone in the pro-choice group advocate for abortion; it may be that most members of both groups would be pleased if abortions would be reduced. The pro-life group seems to consider *life* as the birth of a child, not the quality of life for the child or the affected families. Pro-choice expresses the belief that *freedom* for the prospective parents and their families to make the decision should be paramount.

To the best of my knowledge, no one is proposing any alternative that might 1) reduce the desire for an abortion, 2) reduce abortions more than a prohibition and 3) give the freedom of choice to the affected individuals. Does such an alternative exist?

Consider this:

- The Federal Government passes legislation requiring health insurance companies to cover all birth control;
- The Federal Government commits to covering birth control costs for individuals not covered by insurance;
- The Federal Government also commits to providing sex education for everyone, starting in middle school and continuing throughout their reproductive years. Schools, charitable organizations, and religious organizations would be welcome partners to the education process.

Keeping the formula in mind:

> No unwanted pregnancies = almost no abortions

The alternative might reduce abortions more than prohibition and would be far less expensive for both the insurance companies and the Federal Government.

By now you may well know what I believe about this issue, but I remind you that what I believe is not as important as what you believe. What is needed is **cooperation and respect** by and for all beliefs.

Now is a good time to think, or better yet, meditate about both positions and the alternative, and even to develop alternatives of your own. Use your knowledge to estimate the likely results of each position.

The following two beliefs have influenced me:

- I am influenced by past government prohibitions and the human tendency to violate laws that they do not approve of and to follow the desires of their powerful brain.
- I am also influenced by my belief that people with low incomes or those living in poverty are the most likely to not practice the best birth control methods without financial assistance, and would also be the most likely to seek out dangerous abortions if they felt they needed to break the law and that such violations might also result in otherwise law-abiding citizens becoming criminals.

My thoughts are not as important as what you believe and can visualize. However, I thank you if you read them, and hope that they will provide you with some ideas to meditate on.

When I started to write this book, I decided to avoid any predictions about what would happen if we do nothing to stop the continuous increasing. I wanted to leave those scenarios to the entertainment industry. However, I have recently read a Negative Population Forum Paper written by Greeley Miklashek, M.D. titled *Population Density Stress is Killing Us Now!* His paper provides valuable input into not just what might happen but the reality of what is happening now. It provides input into a population cycle that we might fall into which could lead to our extinction. Until reading his paper I had thought that any one of several horrid things might happen, but it was always in my mind that if our population would crash there would be survivors and the remaining humans would start populating again. Now I am not so sure.

Dr. Miklashek started his medical practice 46 years ago at which time he did not realize that he would find that human over population was going to cause physiological changes responsible for what he now calls the top ten killing "diseases of civilization". Yes, diseases caused by our civilized lifestyle. He blames it on our overactive stress response which is creating abnormally high blood levels of **cortisol**. He further discovered a parallel line of animal crowding research which implicated elevated levels of cortisol. He found that environmental stressors in our daily lives triggered elevated cortisol levels which lead to **anxiety** and **depression**. He found scientific papers showing that cortisol levels could be reduced by medication, relaxation, massage, meditation, Yoga, Tai Chi, music and exercise.

Medical literature dating back to the 20th century demonstrated a connection between cortisol and diseases such as hyperthyroidism, atherosclerosis, heart disease, obesity, diabetes, cancer, suppression of the immune system, increased risk of infection, high blood pressure, kidney disease, peptic ulcers, heart attack, and stroke plus others. He considers these to be diseases of civilization.

He noted that in 1932; 238,851 rural Kenyans and other rural native citizens from remote areas were examined and not a single case of heart disease was found. It was later found that similar rural hunter-gatherers developed the "civilized dieses" within two years of moving to congested urban areas with Western lifestyles.

Dr. Miklashek pointed out that cortisol regulates energy release as well as the immune system. Thus, not only is too much cortisol unhealthy but so is too little. He noted that Population Density stress was making us sick and killing us with a combination, first through diseases of civilization, and then as a result of adrenal fatigue! This is also believed to contribute to the suppression of reproductive functions; which has led me to believe that the extinction of our species is not impossible without an outside catastrophe.

Dr. Miklashek revealed the results of three animal research

studies using rats. The studies showed that when rats were given utopic living conditions, they would increase their population to unsustainable numbers and then when dieback occurred to sustainable numbers the dying did not stop. Despite the utopic conditions and the reduced number of rats the entire population died. This reminded me of what happened to the passenger pigeons when they were taken into captivity and given quality living conditions and food. They never reproduced and just died into extinction. Researchers working with the rat experiment became convinced that rising infertility was the cause of the total die-off.

Today 1 in 6 American couples are unable to have a child after one year of trying; and the same problem is occurring throughout the developed world in urban centers. Dr Miklashek pointed out that the sperm count has fallen by 59% over the past 38 years. He also noted that "The biological evidence of human overpopulation generating population density stress and naturally turning down or even off human reproduction is piling up". He noted that we civilized people are living on the annual 4.3 billion prescriptions prescribed for the "civilized diseases" that both the old folks and the **young** are taking **daily.**

In his forum paper he describes environmental degradation as I have done in previous chapters; and he adds that we are living in what is believed to be the making of the 6th great extinction. He shares my belief by saying "All the biological markers are there for us to see, if we can only find the courage to look the truth in the eyes and take the responsibility for changing the otherwise dismal course of history for our offspring and Mother Earth and all Her creatures great and small." He stresses that we need to do this for the future generation "if there is to be one"!

Since this book is about our population and has concluded that we are presently far in excess of the numbers needed to sustain ourselves and the Earth's wonderful resources, including the other

THE FORBIDDEN SUBJECT

living organisms, at this time, I feel that I must also point out that forcing the fulfillment of unwanted pregnancies would add to our over population.

Your thoughts:

_____
_____
_____
_____
_____
_____
_____
_____
_____
_____
_____
_____
_____
_____
_____
_____
_____
_____
_____
_____
_____
_____

CHAPTER 11

CONCLUSION

There are many quotes in this chapter because great people before us and with us today have envisioned the core of this book and provide wisdom that others have failed to embrace.

Expecting a solution from me? I hope not, because that is a chapter that has yet to be written, and it is not for just me to write. It will be written by **you**, along with the rest of us. However, as a summary, I will share what I believe and encourage you to think deeply about what you believe as you decide what action or actions you are or are not going to take.

Remember, the goal of this book is to change the forbidden subject into the talk of the world! Its purpose is to provide some information to help you to do it in a calm, loving way while appreciating others' views and beliefs. As I have said several times, the opportunity for easy decisions has passed.

I am dividing this chapter into my beliefs, your beliefs, and actions we can take to constructively start reducing the Human population.

MY BELIEFS

First, I believe that planet Earth is unique, and our lives depend on protecting and enhancing its environment.

THE FORBIDDEN SUBJECT

I believe that the planet itself has existed for billions of years and will continue to exist for billions more. The problem is that I also believe that the planet's biosphere is rapidly and unnaturally changing due to our actions. I believe that our changes have caused many plant and animal extinctions and are threating to cause millions more. I believe that our alteration of and damage to the biosphere is accelerating these extinctions and that we may be one of the species to go. Originally; that is when I started writing this book, I thought that our continuing to overpopulate the planet would cause many extinctions but not ours. I thought that we would have a large scale die off but then we would start a new life with our reduced numbers. However, after reading Dr. Miklashek's forum paper I am not so sure.

I believe that those of us who are most able to take corrective actions are living such good lives that they (*maybe that is even you*) have been blind to the daily declines and are afraid that corrective actions won't be popular and will negatively affect their lives. However, I also feel that ignoring them will either lead to disaster or at least make future corrective action less desirable.

I believe that many great environmental organizations exist, and they often accomplish wonderful habitat improvement projects and get some land set aside as "natural" areas, etc. However, this is somewhat like sticking your finger into a hole in a leaking dam. It simply does not correct the basic population problem!

I believe, we must convince the "powers to be" that we are willing to and must accept reductions of our resource consumption in addition to a reduction of our population.

I believe, we must make population control the most important topic on Government's and each of our "plates".

I believe that most people that look at the facts will conclude that continuing the growth of our population will lead to disaster.

I believe the immediate focus needs to stop increasing and to

CONCLUSION

start decreasing our numbers. We must avoid any attempts to develop number goals. We will learn a great deal during the many years required to stop the growth. In the past, looking too far ahead and trying to set numbers has contributed to blurring the focus and derailing needed action. The focus must be clear; the Earth is finite and human population growth cannot go on forever. It should also be clear that we are presently using more natural resources than the Earth and its systems can replenish.

Additionally, I believe that we should share the limited resources with the poor despite such requiring a greater reduction for ourselves. Some things are just plain morally correct!

I believe; actually, I know, that we can be a humane species and live a sustainable life. Happiness does not require excessive consumption; and does not require almost half of our species to be living in poverty.

Remember, the goal of this book is to change the forbidden subject into the talk of the world! Its purpose is to provide some information to help you to do it in a calm, loving way while appreciating others' views and beliefs. As I have said several times, the opportunity for easy decisions has passed. Correction will likely include very unpleasant actions. Looking at how some countries are successfully making reductions may help, but also may be gut-wrenching.

We must realize that if we want to be compassionate and act as **humane** humans, we will need to eliminate the disgraceful poverty that nearly one half of our population struggles with daily; and the annual death of millions of young children due to malnutrition. We must realize that if the impoverished would join the affluent folks, that would greatly increase the consumption of natural resources. Thus, moving one person out of poverty is like adding one person to the planet. But it needs to be done!

THE FORBIDDEN SUBJECT

> *Plans to protect air & water, wilderness & wildlife are in fact plans to protect man."*
> *–Stewart Udall*

    I believe that meditation can help each of us to live happier lives, develop actions, and have better conversations with others when a high level of self-control is needed.

    Recently I heard a short piece about meditation on the radio. It suggested that even fifteen minutes a few times a week could significantly improve our health and well-being. Additionally, I believe that Dr. Doty's information about listening to the heart as well as to the brain is essential. Remembering that the brain is a great rationalizer that is constantly justifying what you desire, it is vital to take the time to calm it down and listen to your heart.

> *"Try to leave the earth a better place than when you arrived."*
> *–Sidney Sheldon*

    I said that we will have time to discuss many of the aspects of stopping population growth and starting reduction. I say this **not** because there is time before damage will occur; that was probably more than a hundred years ago. I say it because just stopping the growth will require significant time and bring about needed reduction will likely require fifty plus years! To make my point I would like to share a simple drastic and unrealistic scenario that I tried. I calculated what population reduction would occur if the world would take a year off from having babies; let's say 2020 for example. Of course, that is not realistic and is impossible because some women are already pregnant and will have their babies in 2020, but that is not important for this exercise, I used it because I could make an educated guess of the population at the end of

CONCLUSION

2019; and the likely deaths for 2020. So, no babies to be born in 2020 and deaths to be like those in 2019. That ought to reduce our population; Yea! it does. But by how much? A little less than 1/2 of 1 %. Recently people were celebrating that our population only grew by about 1% last year. So, we added more than twice as many people last year than we would lose if no babies were born for an entire year. How is that for perspective?

 The organization Negative Population Growth estimates that it would take us about ten years to stop population growth. I could not understand why it would take so long just to stop the growth; but my unrealistic calculation convinced me that they are probably correct. Thus, we will have about ten years of trying different ways of stopping the growth before we will have the opportunity to consider reduction. I can only imagine what the world will be like ten years from now. Think of the vast changes that have occurred in the past ten years and add the ever-increasing advancements that occur each decade and wow! So, we probably have about ten years to change our mind set from growth to what is a sustainable population. Every neighborhood, community, Village, County, State, Nation or other subdivision of people will have the opportunity to inventory their resources, check monitoring records, establish sustainability, and solicit the desired future condition of their citizens. Hopefully, the directives to plan for growth regardless of environmental conditions and citizen desires will be a thing of the past.

 Changing from stopping the growth, to reducing our population to whatever is determined to be a desirable goal based on known conditions will likely require fifty to one hundred years. During that time fantastic advancements of all kinds, accurate data and experience will be available so that knowledge-based goals and actions can be created and then be constantly measured and modified as needed. _Thus, no need to debate them now!_

THE FORBIDDEN SUBJECT

For now, focus on worldwide cooperation and stopping the growth! And don't forget immigration even though it does not change the world population! It is a big part of the problem!

WHAT YOU BELIEVE

It is now time for you to focus what you believe. What will be best for you and the society that you live in. How can you do that without criticizing what others may think or do? What kind of a planet do you want to pass to your children or grandchildren? Yes, the generation that is alive today and will live to experience whatever changes we make or however nature makes the correction; which is likely to be the 6th Great extinction. An extinction that many people believe we are presently living in!

This book is full of information about Earth, our population, nature, and the environment, but what was included is only a speck of the full picture, just as our existence has only been a speck in time. You, however, have personal experiences and knowledge, which provides a good foundation from which to formulate what you truly feel about our ever-growing population. Remember it can be accomplished by any combination of reducing consumption and reducing people.

Before considering what, the world should do; please look at your own life.

Could you reduce your consumption by 50 percent? **Yes 50%!**

Give it a shot, you might start with the car. Could you own a more energy efficient vehicle, electric for example? How about those miles driven; could you live closer to work, ride a bike, ride a motorcycle, carpool, sometimes work from home, not drive the children to all the extracurricular activities and instead do the

CONCLUSION

same things as a family with them at home or at the local park? Just some ideas to think about. Remember, we are shooting for a sustainable society. That can be accomplished by any combination of reducing consumption or reducing people.

The house; how is it doing? Your local energy company would probably do an energy audit for you. Can you save energy and water each day if you tried? As discussed earlier, it would take more than five planet Earths to provide the resources required by humans if all humans consumed at the level of Americans.

Now, what about your property? Is it nature friendly, or is it manicured and requires watering, herbicides, insecticides and fertilizer? Is the yard a mono type of one highbred grass species or a variety of native plants providing habitat for wildlife? Do you know that dandelions are not only beautiful, but the seeds are an important early spring source of food for birds and other animals, and the leaves are good salad greens for you? By the way, they are best if not coated with pesticides! Are you controlling water run-off or contributing to local erosion and pollution?

Once when I was sitting in a waiting room, I overheard two women talking. One woman said how wonderful it was that she was going to be able to fly to California to spend Thanksgiving with her daughter, and the trip was only going to take about 12 hours, doorstep to doorstep. I thought, great; one hundred years ago, she probably could have walked to her daughter's house in less than a half hour. That old fashion trip would have provided some valuable exercise and given her eleven- and one-half hours more to visit, plus saved her daughter a traffic-jammed drive to the airport. Do you have experiences like this?

In general, I have stopped using new information because it occurs so frequently, and it is not the goal of this book to point out every indicator. I am counting on you and the other readers to become observant of the many indicators appearing almost daily in each of our lives, and then to add that information to your own beliefs and use it when you take personal action to discuss the problem. We can convince local, national, and world organizations and leaders that action is necessary.

However, I have recently become aware of a study focused on global warming that ties in directly with your personal efforts to reduce your resource consumption to ease the strain on our planet. Unfortunately, it might alter what you just finished thinking about. Hopefully, it will not cause you to give up on living a more ecofriendly lifestyle, but it will provide you with knowledge about individual consumption.

The study comes from data from Seth Wynes, *The Climate Mitigation*; July 2017;

https://iopscience.iop.org/article/10.1088/1748-9326/aa7541

It includes a chart showing the **tons of carbon dioxide equivalents** that are produced and could be eliminated by taking eleven actions in our personal lives. The study is only applicable to individuals in developed countries, as they are the individuals that are using the most resources.

Wait! Global Warming? This book is about the Human Population not Global Warming. Read on and you will see the astonishing connection between global warming and the human population.

I am going to start with the action that would help reduce global warming the least and work down to the action that would be most beneficial:

Action	CO² Saved
Upgrading lightbulbs to high efficiency	0.1 tons
Recycling	0.2 tons
Hanging cloths outside to dry	0.2 tons
Washing clothes in cold water	0.3 tons
Replacing a standard car with a hybrid	0.5 tons
Eating a plant-based diet	0.8 tons
Switch to an electric car	1.2 tons
Buy green energy	1.5 tons
Avoid one transatlantic roundtrip flight	1.6 tons
Live without a car	2.4 tons
Have one fewer child	58.6 tons

The addition of one child has a tremendous effect: 58.6 tons. The other ten actions only total 8.8 tons, which is less than 1/6 as much as having one fewer child.

Birth control rises to the top again. It is number one in reducing childhood deaths, number one in reducing poverty, number one in protecting endangered species, number one in protecting potable water and below ground aquifers, number one in protecting farmland, number one in protecting our oceans, number one in restoring our forests, and in my opinion number one in eliminating unwanted pregnancies; which might make it number one in eliminating abortions. The list continues, it seems that birth control should be a major part of any planned solution!

> "Being entirely honest with oneself is a good exercise."
> –Sigmund Freud

WHAT ACTIONS CAN WE TAKE?

After thinking about your personal life, think about your community and beyond. There is no need at present to determine what the correct goal for our population should be. What is important now is for our society to realize that a reduction is essential. Slowing down the increase and turning it around to create a decrease is what is presently needed, and it will take several years (probably ten or more) to accomplish. When it does come time to plan for a reduction it must be realized that it too will be a very slow process for practical and demographic reasons. Also, by then we will know much more about existing scientific, environmental, and social conditions so that desired population goals and timetables can be established. Frequent monitoring can keep us on track. Additionally, we must realize that there will never be one "correct" population goal, because each person and their respective societies will have different views on such a figure. Depending on what goal is set at the start it is likely to require 50 to 100 years to accomplish. Once again; plenty of time to monitor and adjust.

Should your brain tell you that you cannot change the world, so forget it and live it up while "you can". Hopefully your heart will kick in and instead encourage you to act.

Should you now be thinking that this was an okay book to read but nothing that you want to get involved in, remember the saying: "If you are not part of the solution you are part of the problem." It is now up to you to develop and carry out corrective actions that you believe will be best for you and the society that you live in, without criticizing what others may think or do. Should you decide that the human population is a problem that simply cannot be ignored, great! If so, you should take the time to focus your beliefs and prepare to discuss the problem more in depth, while constantly being receptive of the responses that you receive from others.

> *"Never never be afraid to do what's right, especially if the well-being of a person or animal is at stake. Societies' punishments are small compared to the wounds we inflict on our soul when we look the other way."*
> *–Attributed to Martin Luther King Jr.*

If your selfish brain says you are just one of 7+ Billion people, what can you do? You may need to remember that so were Rosa Parks, Dr. Martin Luther King Jr., Rachael Carson, Maximillian Kolbe, Nathan Hale, Mahatma Gandhi, Florence Nightingale, and Aleksandr Solzhenitsyn; each of their followers and many other <u>individuals</u> have accomplished great things.

Now; just for the fun of it, let's do some free thinking. This is for the sheer joy of thinking and not to be debated. What might be a sustainable number of humans?

One simple way to calculate this would be to use the calculation that indicated humans started to use more resources than the earth was producing in about 1970. Thus, you could say that is the number to use (3,700,577,650). But that still leaves billions of people living below the poverty level, so we would need to reduce further to allow the poor to consume a reasonable amount of resources for a quality life. How about knocking off a little over one billion more, making the figure about two billion seven hundred million?

Or you could use the statistic that indicated it would take 1.6 planet Earths to provide the resources to satisfy our present use, but again that would not include the very low use by our starving and dying adults and children, or allow them and many others to raise their consumption to approach that of Americans. Thus,

THE FORBIDDEN SUBJECT

we would have to reduce our existing population by at least fifty percent, putting us back into the three billion range and requiring Americans to reduce their usage.

Furthermore, neither of these examples include the concept of restoring wildlife habitats, reforestation, replenishing the aquafers, reducing global warming or cleaning up our air, lakes, rivers and oceans. No matter how you figure it, there is bound to be great disagreement and variance depending on the driving force of our individual brains.

These scenarios are implying a need for a vast reduction; and if debated today would likely reduce our chances of getting moving in the right direction. They can be discussed in the future when our knowledge, experience, and "our world" has changed. **Not now – please!**

What cannot and should not be ignored is the reality that our existing population is too large.

The goal should be to start constructive conversation about the problem. Great accomplishments are often best achieved one step at a time.

Also keep in mind that no other animal on earth can solve our problem, unless it is a disease that we cannot stop. We have the brains that can be accepting of others and the ability to cooperate.

Science can help us to reduce our environmental impact but we must remember that science and all its wonderful advancements has been the main cause of the problem. Past efforts to manage our population often were met with skepticism and the belief that science would develop some solution, but it cannot invent basic resources; they are finite. Science has developed some wonderful things to reduce our impact on the planet. Our farms are producing more food while causing less erosion, our homes are larger and equipped with all kinds of gadgets that use less energy, and our machines are more fuel-efficient. **But**, our population grew too fast and demanded more than science could provide, and the planet is ***finite***.

Our population is not only too large, but almost half of us are living in poverty. Living without the necessities of clean water, adequate shelter, and food. Additionally, we have yet to learn how to get along together. Is this the best that we can do?

I have a couple of dreams for science to accomplish.

First, the easier one. How can we take care of the elderly with a declining population? There won't be enough workers to build the economy and not enough children to care for aging parents. My answer is robotics.

Excuse me for a second; my dog needs out. Robo one would you please let rover out. Thank you and how is dinner coming? Oh! Robo two, don't forget the groceries are to be delivered at three o'clock and will need to be put away. Robo one after letting out the dog would you please make a cup of hot chocolate for Kathy and me.

With the way robotics are advancing today, it is not hard to imagine them solving some of the lost manpower problems.

Secondly, one of our greatest problems is energy production and the environmental problems that it causes. I believe that we are fully capable of solving nuclear fusion. Our existing nuclear plants use ***fission*** not ***fusion***. The sun uses fusion and I believe that we can learn how to do it and produce vast amounts of safe energy.

All this information is great but **what can you and I do** to solve the population problem? I believe that we need to get the attention of our political leaders, our world leading organizations, environmental groups, human rights and medical champions, etc. Get them to join us in changing the forbidden subject from forbidden to the number one priority.

<u>We can do it and using social media can help!</u> Letting everyone on our contact list know how important it is and asking them to spread the word.

Encourage them to read *The Forbidden Subject*, and also read

THE FORBIDDEN SUBJECT

Better not Bigger, by Eben Fodor. Go to your local political and planning commission meetings and request that they read both books.

When you receive solicitations from charities that you support and even the ones that you don't; find out who their president is and send them a personal note or letter requesting that they recognize that their goals cannot be achieved without or at least would be better achieved with population control. Encourage them to recognize such in their solicitations and publications; and to read this book. Recommend that they include population control articles in their publications.

I know that the above are requests to help sell this book and doing such is very unusual, but this is not being written as a means of earning money but is intended to fulfill a driving passion that I have to do something about before my leaving this planet. Afterall, spreading the word is the goal of the book! Additionally, I intend to direct at least half of any royalties received toward further spreading the word and to charities that focus on population control, women's rights and the environment.

Encourage local Governments to resist growth pressures from higher units of Government; and instead direct their employees and existing planning commissions to solicit neighborhood groups to execute true planning in place of existing growth planning. These newly focused groups can look at existing conditions, environmental limitations, social values, industrial conditions, infrastructure, public services, schools, food production, wildlife needs and values, etc. Such planning can build the foundation for good national and planet earth planning.

Each of us should be able to make the important decisions for our own lives. As for me, I am happy to live if I can continue to benefit and socialize with man, beast, and all of nature; when I cannot, please let me die!

CONCLUSION

So, I am not providing a solution, only a challenge. It is a challenge that needs to start with constructive, loving **conversation**. It is not forbidden. **_We can and must do it!_**

"Anything that's human is mentionable and anything that is mentionable can be manageable."
−Fred Rogers

So, let's get talking and acting on our beliefs!

"The world is my Country, all mankind are my brethren, and to do good is my religion."
−Thomas Paine

"Mine, too."
−Morse Reese

135

THE FORBIDDEN SUBJECT

What do you believe/What are you going to do?

ACKNOWLEDGMENTS

If I would recognize all the people who have helped me or have put up with me while I have been writing this book, I would probably have created another book. So, all that I will say to them is *thank you!*

However, I must specifically recognize some of the folks who have been the driving force to get me started and to get me re-started when I was in one of many slumps.

First and foremost is Dick Yost and his wife Janice (my cousin); who inspired this non-writer, non-author to do it anyway! Without Dick I never would have started.

Next is my brother Harry Reese "Tim" and his wife Kate who provided me with solitude, criticism and sanctuary in their home. The solitude provided what I needed to focus, and kick start the project.

Of course, it was my wife Kathy who did much of the grammatical editing and put up with my frustration and ranting about this subject due to my overwhelming love of nature and distain of our treatment of the environment.

I would also like to thank Salina Swancer for help with the cover concept to define the book; my daughter Jenny and her husband Sam for their computer assistance; and my sons Tim and Tony who provided reviews and improvements. Cousin Jean Siebecker's "teachers" close eye review also contributed and is deeply appreciated.

Those of you that helped and were not mentioned, please know that you are appreciated and I hope that the finished product will make you feel that your help was worthwhile.

CREDITS

We live in a new age and most of my credits are simple research on the internet using several encyclopedias, Wikipedia, and even Merriam Webster's Collegiate Dictionary which has good coverage of subjects like geologic time. I also used the *World Atlas of Geology* and the following books and papers: Specific studies, quotes and other information are credited in the text.

a) The Case for Fewer People by Negative Population Growth Inc. and several of their forum papers which are specifically noted in the text. (A1)
b) The population *Connection* Magazine and it's wonderful President's Notes.
c) *Remembering: Voices of the Holocaust* by Lyn Smith. (A2)
d) *Into the Magic Shop*, by James R. Doty, MD, which provided guidance for meditating and insight into how we humans decide to do what we do.
e) *Silent Spring* by Rachael Carson which provided inspiration for me and hope for the future. She had the advantage of being an accomplished author and having a subject that people were anxious to embrace.
f) *Better Not Bigger* by Eben Fodor which provided excellent information about community planning and should be "repeatedly" read by all public servants.

(A1) – I highly recommend joining them as I believe their work and forum papers to be well thought out and balanced.

(A2) – Their focus is pointing out serious problems that are directly related to our overpopulation. An example is – While global warming is a tremendous castrophy and is caused by carbon emissions the real problem is the overpopulation that is creating the emissions. They also provide education programs for school and other groups.